NATHAN COPPEDGE'S PRACTICAL PERPETUAL MOTION HANDBOOK

PRACTICAL PERPETUAL MOTION HANDBOOK

Famous Perpetual Motion Paper, **p. 5**

Sample Stats, **p. 6**

Stats & Specifications, **p. 7**

Tips on Perpetual Motion, **p. 12**

- **CAREFUL INSTRUCTIONS,**
    - Underwater Bubblevator, p. 14
    - Vertical Leverage Device, p. 15
    - 1st Fully Provable, p. 18
    - Cat-Trap Device, p. 21
    - Swivel Lever Device, p. 24
    - Escher Machine, p. 26
    - Bubblevator, p. 30
    - Modular Buoy Device, p.33
    - Underwater Escher Machine, p. 36
    - Unnatural Torque, p. 38

Inventor's Journal, p. 39

BIO,………………………………… p. 182

Updated 2023.

NATHAN COPPEDGE'S PRACTICAL PERPETUAL MOTION HANDBOOK

© 2000, '05, '06, '07, '08, '09, 2010, '11, '12, '13, '14, '15, '16, '17, '18, '19, '20, '21, '22 Nathan Coppedge. Some material released on Quora.com reusable with proper citation of the author.

NATHAN COPPEDGE'S PRACTICAL PERPETUAL MOTION HANDBOOK

NATHAN COPPEDGE'S PRACTICAL PERPETUAL MOTION HANDBOOK

(...)

NATHAN COPPEDGE'S PRACTICAL PERPETUAL MOTION HANDBOOK

PERPETUAL MOTION PAPER
USED FOR CALCULATING WORKING PERPETUAL MOTION PROPERTIES

PERPETUAL MOTION CHEATSHEET
v 1

>H	>V
MECHANICS	
COMBOS	1-PRINCIPLE
GRAVITY	CYCLICAL
FUEL	I/O
UNIDIR	DUAL-DIR
STEMS	MOBILE U
SUBCYCLES OPTIONAL	
TENSILE	BENDY
L. STRUCT	H. STRUCT

SOMETIMES...

LEVER	WHEEL ONLY
HEAVY CW	LIGHT CW
SYMMETRIC	

[nathan coppedge @ Quora.com]

HORIZONTAL LEVERAGE DEVICES

MAX LEVERAGE = _____

MIN LEVERAGE = _____

MAX COUNTERWEIGHT MASS = MIN LVG + 1 = _____

MIN COUNTERWEIGHT MASS = (MAX LVG / 2) + 1 _____

ADDITIONAL LONG-END MASS = 1 (CONSTANT)

OVER-UNITY =

[(MAX CTRWEIGHT - MIN CTRWEIGHT) / MAX LVG] + 1 MASS X 100

= _____ % OU + 100 IF FLYING

UNUSUAL CASES (FOR ESCHER PUT EFF = 1.25)

AMOUNT OF EFFICIENCY (MASS X LEVERAGE) _____

(X) GRADIENT IN PERCENT DEGREES (0 -1) _____

=_____ = E

AMOUNT OF RESISTANCE (MASS X LEVERAGE) _____

(X) GRADIENT IN PERCENT DEGREES (0 - 1) _____

=_____ = R

(E / R) X 100 = _____ % OU

SAMPLE MATHEMATICS

MACHINE	LVG	MIN*	MAX*
RL 4.2	5 : 1	>3.5	<6
IVL	4 : 1	>3	<5
SWVL	2 : 1	>2	<3
1STFP	1.75-2.25:1	>2.125	<2.75
ESCH	1 : 0.5	BALL 1X NO CW	
FLYING	Same,	Mass now= Byncy	

PERPETUAL MOTION EQUATIONS

PERPETUAL MOTION:
- GENERALIZED PERPETUAL MOTION: Results = Eff (1,2,3...) + Difference
- Min Heavier Mass = (Max Lvg / 2) + 1
- Max Heavier Mass = Min Lvg + 1
- Min Lvg = Max Heavier Mass - 1
- Max Lvg = (Min Heavier Mass - 1) X 2
- Over-Unity = Heavier Mass Rng / Lvg Ratio + 1 X 100 (%)
- Smaller Mass = 1X

FLYING MACHINES, ETC:

- 6.5 + the Function Number yields the objective element number for purposes of technology in the Function Spectrum. E.g. a Function Number of 1.5 yields an element number of 8. The typical functional perpetual motion machine has a refined element number of 7.5 if non-flying, and 8.5 if flying. —Perpetual Motion Object Language
- PM Cars Extra Mass < OU - 100%
- Flying Vehicles Extra Mass < OU - 200%
- Flying does not work when Lvg Rng >= 1/2 max leverage.

- Flying Machines Window = Max Larger Buoyancy - Min Larger Buoyancy
- Flying Max Larger Buoyancy = (Min Lvg) additional mass cancels with 1 unit buoyancy
- Flying Min Larger Buoyancy = (Max Lvg / 2) additional mass cancels with 1 unit buoyancy
- Flying OU = Larger Buoyancy Range / Leverage Ratio + 1 * 100 (%) + 100 for buoyancy.
- Flying Smaller Buoyancy = 1X
- Secret of perpetual motion flying machines: <u>Improved Balancing Balloons Theory</u>
- Planetoids: Estimated < (Phi / 2 + 1 * 100 =) 180.9% OU,
- Max sustainable mass resistance to Earth perpetual motion = <<0.809 X distance (Earth diameters). With an estimate saying Earth's max output is about 110% with rotation.
- The Max Min required distance to resist the Sun with PMMs is 4726 AU assuming 110% OU
- Perpetual motion holds the key to the material world.
- Flying machines hold the key to the universe.

8

ADDITIONAL SPECIFICATIONS:

Cost Effectiveness: Approaching ideal depending on management, approximately > $1,000,000 / MW

First Evidence: First major partial evidence in September 2000, and Nov 9 - 10, 2013. Some minor principles from earlier.

Major Types: Perhaps five or so, to several dozens, many variations on two types mentioned below.

Applications: Numerous applications such as enhanced bicycles and self-powered batteries, and magical books depend on building a fully-working model first.

Ratings: Ratings may fall between >100% to <150% which means >1X to <1.5X unity (expressed as gravities of 1X mass minus resistance, and not always accounting for gradient as gradient may vary).

Principle of Operation: Typically (1) Leverage vs. short-distance counterweight combined with movement of ball weight on-off slope of around 10 degrees or less, and some geometric considerations OR (2) Large ball at short distance counterweighted by a light counterweight at much longer distance, with similar considerations.

Equation: [New Unified Equation]: "Heavier mass > 50% of opposing lvg plus un-weighted difference and < 100% of opposing leverage plus un-weighted difference."(—The Theory of Everything) Note this is only an approximation, some devices may not use leverage, but may instead use wedges or pulleys, which may or may not be directly comparable. For example, a wedge is thought to have a 125% effect (?) and not all pulleys function like a balance.

Maximum recommended leverage for a perpetual motion machine: 1X to <4X measured in lengths of the shorter end.

Maximum recommended heavier mass: 1X to <5X lighter weight.

Maximum Gradient (Steepness of slope): <22.5 degrees often less.

Elements (D + 2) = 5 (Ex, lever, ball, counterweight, pivot, fixed track OR ball, slope, backboard, precise angle, drop).

(Actual height / Recoverable loss of height) X 1.5 = distance traveled in proportions of the short end.

For example, 1X height / 0.25 loss X 1.5 = 6X distance, confirmed in Improved Anachronistic

1 / 0.333 X 1.5 = 4.5 distance, confirmed with NIBW6.

TIPS ON PERPETUAL MOTION

People say perpetual motion inventors are either tricksters or fools. I partly agree.

You might say there are at least two other categories: Lucky Fools, and Unwilling Tricksters.

Although it is unlikely a genius would try one given the tradition of failure, sometimes a foolish genius or Lucky Fool might make progress simply by doing something no one else had tried.

Also, there are unwilling tricksters who begin by attempting something they think is authentic but end by deceiving their audience to cover up their feelings of failure, or because they are in denial about their specific design's weaknesses.

In this case the trickster could be less intelligent than a lucky fool because they don't even know how to be honest with themselves, meanwhile a conventional trickster might not even try.

I welcome others to do more precise experiments if they find the impulse.

NATHAN COPPEDGE'S PRACTICAL PERPETUAL MOTION HANDBOOK

CAREFUL INSTRUCTIONS ON BUILDING PERPETUAL MOTION

UNDERWATER BUBBLEVATOR

PROBABILITY OF DEFINITE SUCCESS: MEDIUM-HIGH

PRODUCT-WORTHINESS: VERY HIGH (SOME CONSIDERATIONS)

EXTRA CONSIDERATIONS: SINCE THE DEVICE IS MEANT TO RUN UNDERWATER, IT COULD POTENTIAL RUN WITH SLIGHTLY REDUCED EFFICIENCY WITHOUT WATERTIGHT SEAL PROBLEMS. THIS MAKES IT AN APPEALING, EASY-TO-MANUFACTURE PRODUCT AS SOON AS A WORKING MODEL IS DEMONSTRATED. FEW IF ANY EXPERIMENTS HAVE BEEN DONE TO CONFIRM OR DENY IT AS OF JANUARY 2022, OTHER THAN THE MORE DIFFCULT SEEMING STANDARD LAND-BASED BUBBLEVATOR EXPERIMENT.

THE UNDERWATER BUBBLEVATOR

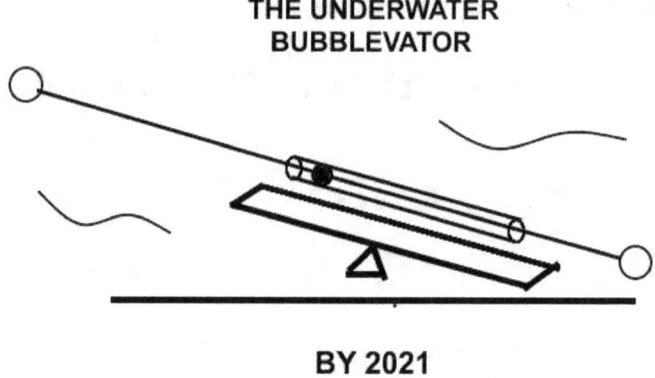

BY 2021

First known diagram: November 11, 2021

Rating: < 116.67 % Conventional OU

"Spaghetti" Leverage: 1 to 1.5 : 1

Center Mass Range: >2.75X <3X

Additional "spaghetti" buoyancy: 2

Estimated Maximum gradient: < 7.425 degrees.

Standard Equation: Min mass is equal to the Max Lvg / 2 + difference. Max mass is equal to the Min Lvg + difference.

...

VERTICAL LEVER DEVICE

PROBABILITY OF DEFINITE SUCCESS: MEDIUM-HIGH (WITH SOME CONSIDERATIONS)

PRODUCT-WORTHINESS: HIGH (LOW REAL-ESTATE CROSS-SECTION, HIGHLY PORTABLE)

EXTRA CONSIDERATIONS: AS FAR AS LIKELY SUCCESS, ONLY THE 4:1 MODEL IN A SPECIFIC ARRANGEMENT SEEMS TO BE FULLY WORKING, AND THIS DESIGN WILL REQUIRE PERSISTANCE BEFORE PROGRESS IS MADE, WHICH IS ONE OF THE REASONS PROBABILITY WAS MARKED HIGH. THIS DESIGN REQUIRES PROFESSIONAL ATTENTION BEFORE THE PROBABILITY BECOMES HIGH. HOWEVER, THE 4:1 MODEL HAS AN EXTREME REQUIREMENT FOR HYPER-LIGHTWEIGHT HIGH-TENSILE STRENGTH PARTS AS FAR AS THE STRUCTURE OF THE LEVER. FORTUNATELY, IF THE MASS OF THE BALL AND COUNTERWEIGHT ARE SCALED UP WITH APPROPRIATE TRACK STRUCTURE, CONVENTIONAL MATERIALS LIKE STEEL AND WOOD MIGHT BE USED THOUGH STILL WITH A REQUIREMENT FOR REMAINING AS LIGHTWEIGHT AS POSSIBLE. THE BALL AND COUNTERWEIGHT NEED NOT BE LIGHTWEIGHT BUT IT RAISES STRUCTURAL CONCERNS. IT IS MY BELIEF THAT SOMEONE CAN RESOLVE THIS. CRUDE EXPERIMENTATION WITH WOOD DID NOT SHOW GOOD RESULTS, BUT

PLASTIC AND ALUMINUM SEEM TO WORK WELL. HEAVIER MATERIALS COULD BE USED WHEN STRUCTURE IS TALLER THAN 40 INCHES, BUT ONLY WITH DIRE CAUTION. FRICTION IS ALSO A HUGE CONCERN, BUT IN THE BEST CASES IT CAN BE SIGNIFICANTLY REDUCED. USE OF MARBLES OR BALL BEARINGS LARGER THAN 0.5" IS RECOMMENDED AND THIS CAN CREATE STRUCTURAL CONSIDERATIONS.

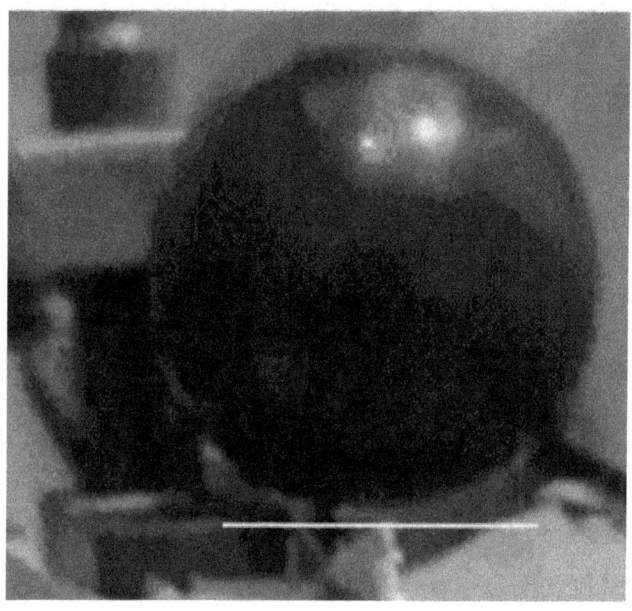

ABOVE: A 3 : 1 VERSION OF THE VERTICAL LEVER DEVICE BASICALLY WORKING (HARDER WITH 3: 1 THAN 4: 1 BUT FEWER STRUCTURAL CONSIDERATIONS)

NATHAN COPPEDGE'S PRACTICAL PERPETUAL MOTION HANDBOOK

PROOF OF THE IMPROVED VERTICAL LEVER (IVL) PERPETUAL MOTION CONCEPT, ONE OF THE FIRST THEORETICALLY-SUPPORTED PERPETUAL MOTION MACHINES (OF THE 1ST KIND).

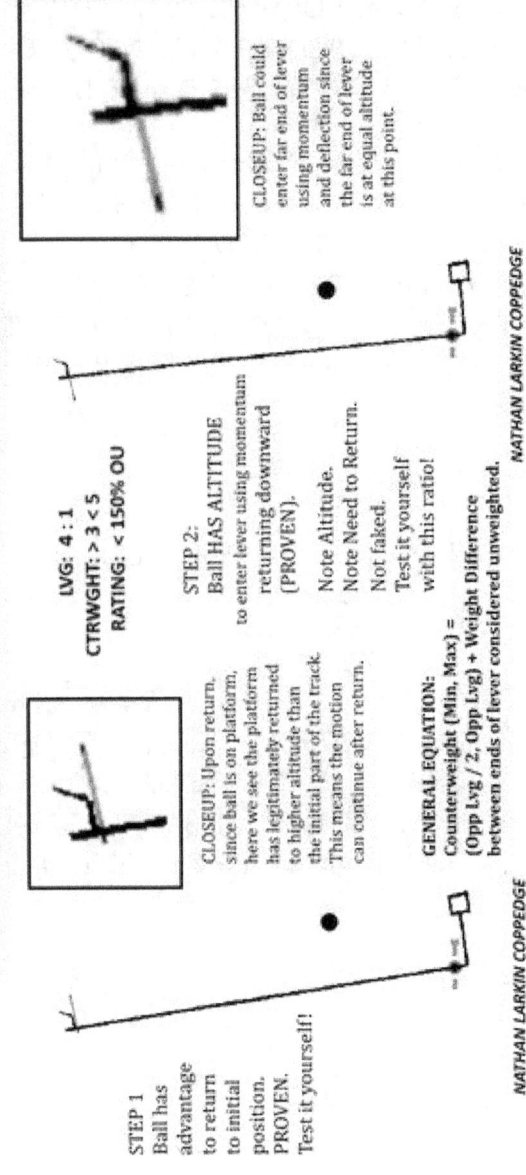

LVG: 4 : 1
CTRWGHT: > 3 < 5
RATING: < 150% OU

STEP 1
Ball has advantage to return to initial position.
PROVEN.
Test it yourself!

CLOSEUP: Upon return, since ball is on platform, here we see the platform has legitimately returned to higher altitude than the initial part of the track. This means the motion can continue after return.

STEP 2:
Ball HAS ALTITUDE to enter lever using momentum returning downward (PROVEN).
Note Altitude.
No Need to Return.
Not faked.
Test it yourself with this ratio!

CLOSEUP: Ball could enter far end of lever using momentum and deflection since the far end of lever is at equal altitude at this point.

GENERAL EQUATION:
Counterweight (Min, Max) = (Opp Lvg / 2, Opp Lvg) + Weight Difference between ends of lever considered unweighted.

NATHAN LARKIN COPPEDGE

NATHAN LARKIN COPPEDGE

ABOVE: HERE IS THE 4: 1 OPTIMIZED. NOTE: COUNTERWEIGHT WOULD BE POSITIONED STRAIGHT OUT FROM LEVER UNLIKE IN THIS DIAGRAM.

INSTRUCTIONS: BUILD AS IN THE DIAGRAM, BUT WITH A 3RD DIMENSION (DEPTH) IN WHICH ON ONE SIDE THE BALL RISES SUPPORTED AND ON THE OTHER SIDE THE BALL FREE-FALLS SUPPORTED ONLY BY THE LEVER. YOU MAY NEED TO INNOVATE WITH DEFLECTION TECHNIQUES AND ADDITIONAL ANGLES TO MAKE IT WORK.

1ST FULLY PROVABLE

PROBABILITY OF DEFINITE SUCCESS: MEDIUM

PRODUCT-WORTHINESS: VERY HIGH

EXTRA CONSIDERATIONS: THIS DEVICE WAS ONE OF THE FIRST TO SHOW MATHEMATICAL APTITUDES. THE PRECISE AMOUNT OF COUNTERWEIGHT RELATIVE TO THE WEIGHT OF THE BALL IS VERY IMPORTANT WHICH IS NORMAL FOR PERPETUAL MOTION.

NATHAN COPPEDGE'S PRACTICAL PERPETUAL MOTION HANDBOOK

WHAT I CALL THE "1ST FULLY PROVABLE PERPETUAL MOTION MACHINE"
[MODIFIED FOR GREATER WORKABILITY]

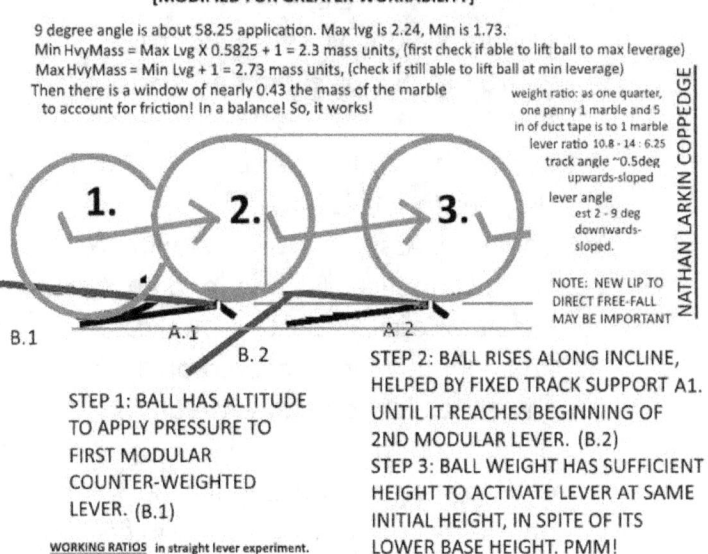

STEP 1: BALL HAS ALTITUDE TO APPLY PRESSURE TO FIRST MODULAR COUNTER-WEIGHTED LEVER. (B.1)

WORKING RATIOS in straight lever experiment.

STEP 2: BALL RISES ALONG INCLINE, HELPED BY FIXED TRACK SUPPORT A1. UNTIL IT REACHES BEGINNING OF 2ND MODULAR LEVER. (B.2)

STEP 3: BALL WEIGHT HAS SUFFICIENT HEIGHT TO ACTIVATE LEVER AT SAME INITIAL HEIGHT, IN SPITE OF ITS LOWER BASE HEIGHT. PMM!

INSTRUCTIONS: FOLLOW MATH AND DIAGRAM CLOSELY. REMEMBER THERE ARE COUNTERWEIGHTS ON THE SHORT ENDS OF THE LEVERS. THE LEVERS ARE REPEATED. IT IS A MODULAR DEVICE WHICH COULD HAVE A LONGER OR SHORTER LOOP DEPENDING ON THE NUMBER OF MODULES.

- Date of Invention: July 12, 2016.
- State of affairs: Strong suggestion of workability. Not easy due to modular construction and in some cases need for extremely heavy weights or perfect hollow metal spheres.

- <131% conventional Over-Unity.
- Leverage: 1.75-2.25: 1
- Counterweight Mass: >2.125 to <2.75X (previous estimate 1.8X to 2.4X)
- Maximum Gradient: Approx < 13.05 degrees (not calculated).
- Equation: Assuming ball = 1 with variable application, and long end has additional 1 constant application, and counterweight located on shorter end, and counterweight is designed to direct ball on opposite end up slight supporting incline before ball applies leverage, Unified Counterweight Mass Formula = Min Lvg + 1 > (Max Lvg / 2) + 1.

CAT-TRAP APPARATUS:

PROBABILITY OF DEFINITE SUCCESS: MEDIUM (RELATIVELY HEAVY COUNTERWEIGHTS WITH LIGHTWEIGHT BALLS IS THOUGH TO INCREASE PROBABILITY)

PRODUCT-WORTHINESS: HIGH

EXTRA CONSIDERATIONS: IMPORTANT TO DESCRIBE OPERATION WHEN SELLING PRODUCT, IF PRODUCT WORKS. PEOPLE MAY NOT UNDERSTAND IT WORKS UNLESS YOU CREDIT NATHAN COPPEDGE. ALSO, THIS DEVICE HAS A HIGH RISK OF BEING DESTROYED DURING DEMONSTRATIONS DUE TO SUSPICIONS THAT IT IS RUN ELECTRONICALLY. THOUGH IF BUILT CORRECTLY IT IS NOT RUN ELECTRONICALLY. THIS DESIGN MAY ACTUALLY BENEFIT BY USING A SMALL BALL THAT JUST SKIMS ALONG USING HEAVY COUNTERWEIGHTS, WHICH IS A PRINCIPLE THOUGHT OF BY JORGE VARGAS. ALL PUBLISHED EXPERIMENTS AND DIAGRAMS SO FAR ARE BY NATHAN COPPEDGE, AS NATHAN COPPEDGE WAS NOT SEEKING A PATENT.

ABOVE: AN EARLY DEMONSTRATION OF THE BASIC PRINCIPLE

...

INSTRUCTIONS: Build a very shallow lever, in about a 1.5: 1 ratio, with the far end having a slightly inward-angled lightweight board. Build several of these in loop, with each unit having a deflection just above the hinge of the lever. Each lever will have a >1.75X to <2.5X counterweight, measured in ratio to a 1X ball which is meant to follow the chain of levers, possibly using sidewalls to prevent the ball from escaping. The balls must be tall and lightweight but with the correct amount of mass, and the levers must hinge solidly and with correct blocking at the top and bottom of motion. The counterweights must also be solidly attached with the correct amount of mass of about 2X the mass of the ball or perhaps slightly more depending on the structural mass of the lever. Remember, the levers must be very shallow so that the momentum combined with the relatively high altitude of the midpoint of the ball allows the ball to press on the next shallow upward-and-downward hinged lever, in this case moving inwards from the tip along each lever and deflecting. The energy from the counterweight because of the counterweight's mass, combined with the angle of the panel at the beginning of the lever is designed to keep the ball moving and even accelerating up to a certain speed, as the ball is directed by the deflecting panels along the course around the levers. Since this is designed to function like a perpetual motion machine, if built successfully it could be worth $1,000,000.

SWIVEL LEVER DEVICE

PROBABILITY OF DEFINITE SUCCESS: MEDIUM

PRODUCT-WORTHINESS: MEDIUM

EXTRA CONSIDERATIONS; THE BIGGEST CONCERN IS NOT TO CONFUSE BETWEEN PARTIAL AND COMPLETE MODELS. A COMPLETE MODEL COULD BE A VALUABLE PRODUCT, BUT SELLING PARTIAL MODELS BEFOREHAND COULD RADICALLY REDUCE SALE PRICE. OVERALL, THIS DESIGN IS NOT AS MARKETABLE AS SOME OTHER MODELS HOWEVER, IF SOLD ALONG WITH OTHER MODELS THE APPEAL MAY GROW CONSIDERABLY. THIS DESIGN IS NOT AS PORTABLE AS OTHER DESIGNS AND CAN BE COUNTER-INTUITIVE TO ASSEMBLE, HOWEVER, THERE IS A KIND OF NIFTINESS TO IT'S OPERATION IF YOU CAN GET IT WORKING.

INSTRUCTIONS: A counterweight is placed on the short end of a lever, with the long end in a 2:1 ratio with the short end. A small basket is placed slightly inside the tip of the long end, so that when a ball is moved very slightly diagonally upwards by the counterweight, the ball is deflected by a panel or wedge inwards towards the basket. When the counterweight is approximately >2X - < 3X the mass of the ball (or in practice very close to 2.5X to slightly more), then once the ball reaches the basket it will lift the counterweight as the ball moves downwards diagonally with the lever following the same path backwards. With enough energy, if the track is shallow enough, the ball may be deflected, perhaps due to only slight inward and outward motion, to return to the beginning of the track, where the same process will (or is designed to) repeat. These are designed to work as a perpetual motion machine as one unit, and thus might sell for $1,000,000 if the loop repeats.

ORIGINAL / REVERSE ESCHER MACHINE (COPPEDGE)

PROBABILITY OF DEFINITE SUCCESS: HIGH

PRODUCT-WORTHINESS: VERY HIGH

EXTRA CONSIDERATIONS: HOW TO MARKET PRODUCT SO AS NOT TO UNDER-HYPE (MAY HELP TO GET LITERATURE AND DIAGRAMS FROM NATHAN COPPEDGE IF YOU KNOW HOW TO BUILD IT). PLEASE FOLLOW INSTRUCTIONS CAREFULLY TO BUILD CORRECTLY. THIS DESIGN, LIKE THE VERTICAL LEVER, REQUIRES A LOT OF TRIAL-AND-ERROR TO GET CORRECT. THE KEY HERE WOULD BE TO LEARN HOW

TO MASS MANUFACTURE IT WITH PRECISION TECHNIQUES SO AS TO AVOID LABORIOUSLY MEASURING THINGS EACH TIME. IF YOU CAN MASS PRODUCE IT AS A SINGLE UNIT OR AS A REPEATED CONNECTABLE MODULE, IT WILL HAVE MORE POTENTIAL AS A PRODUCT.

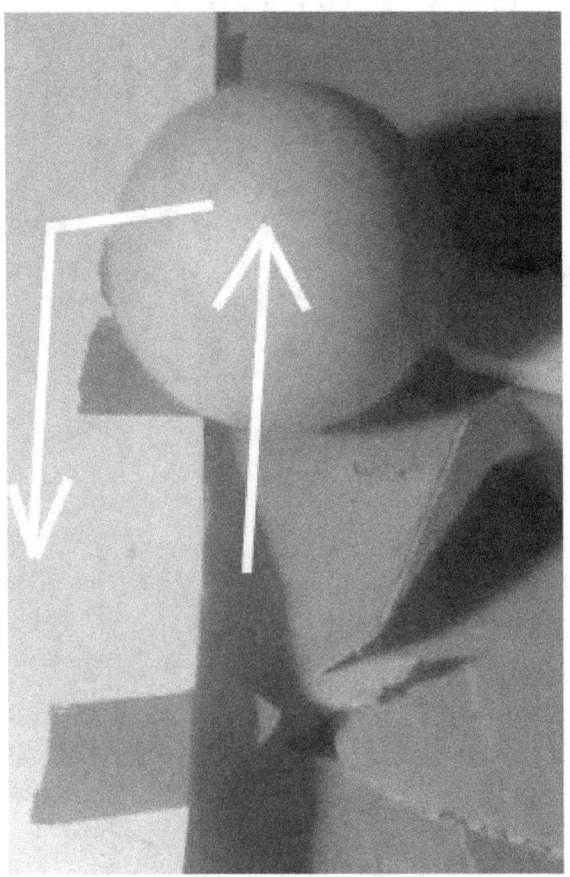

NATHAN COPPEDGE'S PRACTICAL PERPETUAL MOTION HANDBOOK

THE ESCHER MACHINE

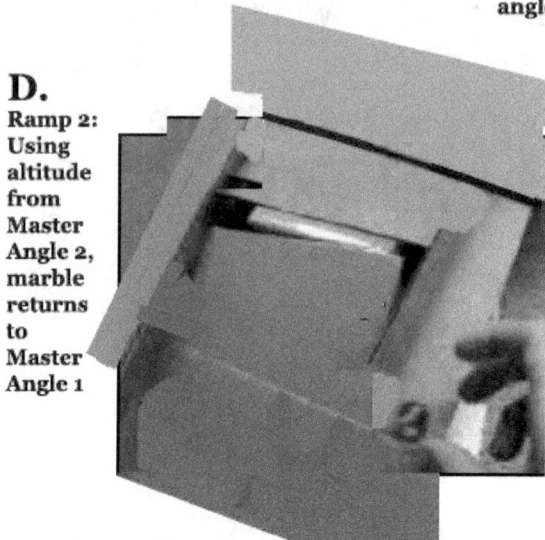

C. Master Angle 2: marble rolls upwards again, using a differently-directed master angle

D. Ramp 2: Using altitude from Master Angle 2, marble returns to Master Angle 1

B. Ramp 1: A downwards motion is possible due to the gain in height

A. Master Angle 1: marble rolls upwards using a horizontal slope

NATHAN COPPEDGE

INSTRUCTIONS: Build a specific angle directed upwards at about 1.1 degrees serving as a wedge, with a ball positioned over the wedge rolling against a backboard angled at about 45 degrees, with the rail for the wedge angled a particular way on the board, and the board rotated slightly in such a way where if the board is square, the far side is elevated significantly, but the near side is also elevated slightly but not as much, and the near side of the board at the top is angled upwards compared to the far side of the top of the board so far as is possible. If you can figure out how to build these modularly, it is designed to be a perpetual motion machine. These might sell for $1,000,000 each if they work to loop the ball on it's own power in a circle.

...

JEN'S BUBBLEVATOR

PROBABILITY OF DEFINITE SUCCESS: LOW

PRODUCT-WORTHINESS: MEDIUM (SOME CONSIDERATIONS)

EXTRA CONSIDERATIONS: JEN'S BUBBLEVATOR IS A WATER-POWERED DEVICE WHICH DOES NOT HAVE AS MUCH EXPERIMENTAL SUPPORT AS SOME OF THE OTHER MODELS. THERE MAY BE A LOT OF FINICKY CONSIDERATIONS HAVING TO DO WITH REDESIGN AND KEEPIING THE DEVICE WATERTIGHT IF NECESSARY. ONE ADVANTAGE IS THIS VERSION OF THE DEVICE COULD IN PRINCIPLE OPERATE ON LAND WITHOUT A LARGE AMOUNT OF WATER, HOWEVER, WATER SEALS ARE NOTORIOUSLY DIFFICULT IN CONSUMER PRODUCTS AND THIS MAY MAKE THE DEVICE PRONE TO BREAKING OR CLOGGING. I LISTED IT HERE BECAUSE IT MAY NOT BE IMPOSSIBLE, AND IF BUILT TO PERFECTION IT MIGHT DEMONSTRATE SOMETHING UNUSUALLY CLEVER. SOME INNOVATION IS REQUIRED AND THIS MIGHT BECOME A HIGHLY SUCCESSFUL PRODUCT. ALSO BE CAREFUL TO USE THE CORRECT MATH, SOME VERSIONS OF THE MATH DO NOT WORK, BUT OTHERS DO SEEM TO WORK. MAKE SURE THE BUBBLE HAS ENOUGH FORCE TO MOVE THE 'SPAGHETTI STRINGS' TO DO THIS KEEP THE DEVICE SWAYING AT CLOSE TO LEVEL, BUT NOT QUITE.

JEN'S BUBBLEVATOR

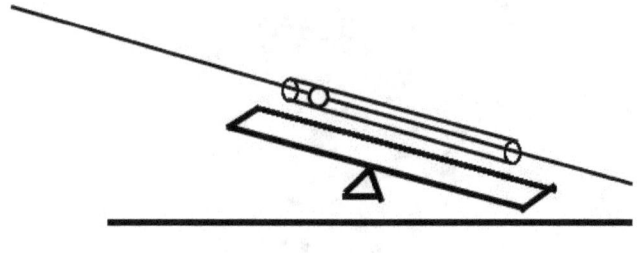

BY 2015

Rating: < 116.67 % Conventional OU

"Spaghetti" Leverage: 1 to 1.5 : 1

Buoyancy Range: >2.75X <3X

Additional "spaghetti" mass: 2X

Estimated Maximum gradient: < 7.425 degrees.

Standard Equation: Min buoyancy is equal to the Max leverage / 2 + the difference. Max buoyancy is equal to the min leverage + difference. Min larger mass is equal to 2X the leverage / 2. Max larger mass is equal to the min leverage. The counterweight, since it is smaller due to meeting reduced resistance, is assumed to be equal to 1. The ball mass is variable because it is the larger mass.

MODULAR BUOY DEVICE

PROBABILITY OF DEFINITE SUCCESS: LOW

PRODUCT-WORTHINESS: MEDIUM

EXTRA CONSIDERATIONS: THIS DEVICE POSES SOME CHALLENGES DUE TO BEING BUILT UNDERWATER. THE WATER RESISTANCE IS KNOWN TO AFFECT THE MOVEMENT OF THE BUOYS IN A DETRIMENTAL WAY. HOWEVER, ADOPTING A LOWER STANDARD THAN NORMAL FOR PERPETUAL MOTION THE DEVICE STILL APPEARS OPERABLE AND IS AN INTERESTING EXAMPLE OF UNDERWATER LEVERAGE. THIS DEVICE MAY ULTIMATELY SUFFER FROM MARKETING ISSUES IN PART DUE TO ITS UGLY APPEARANCE BUT IS OTHERWISE INTERESTING AND MAY LOOK ATTRACTIVE IN A LONG UNDERWATER SERIES WHICH COULD BE OPERATED BY MORE THAN ONE BUOYANT SPHERE IF STAGGERED CORRECTLY. IT COULD BE BILLED AS ONE OF THE MOST ATLANTEAN DEVICES EVER CONCEIVED. IF THEY ONLY HAD PERPETUAL MOTION THEY MIGHT HAVE SURVIVED!

NATHAN COPPEDGE'S PRACTICAL PERPETUAL MOTION HANDBOOK

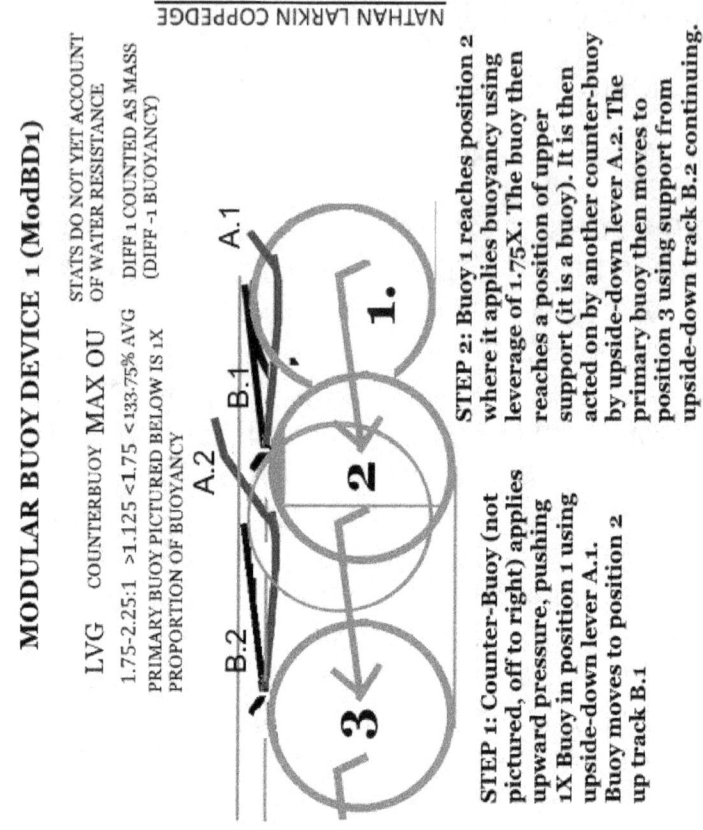

INSTRUCTIONS: BUILD LIKE THE 1ST FULLY PROVABLE, EXCEPT UPSIDE-DOWN, USING SPECIFIC NEW VALUES FOR THE COUNTERBUOY AND MOBILE BUOY IN PLACE OF THE WEIGHTS IN THE ORIGINAL DEVICE.

Date of Invention: July 19, 2021

Precedents: 1stFPPMM, August 2016

Rating: 133.75% Conventional OU

Leverage: 1.75 to 2.25 : 1

Counterbuoy Buoyancy: > 1.125 < 1.75

Difference: -1 (mass on long end counts towards counterbuoy buoyancy)

Additional factors: long end of each lever is assumed to have exactly 1X additional effective leverage coming from structural mass of the lever itself. This means the lever is lightweight but opposing the primary buoy, and helping the counterbuoy.

UNDER WATER ESCHER MACHINE

PROBABILITY OF DEFINITE SUCCESS: VERY LOW

PRODUCT-WORTHINESS: HIGH (SOME CONSIDERATIONS)

EXTRA CONSIDERATIONS: ONE LIMITATION OF THIS DEVICE IS IT MUST OPERATE UNDER WATER. THEREFORE, IT IS LESS LIKELY THAT THE DEVICE WILL BE DIRECLTY VIEWABLE UNLESS IT IS SET UP IN A SPECIAL UNDER WATER DISPLAY. THE UNDER WATER QUALITY CAN GIVE THIS DEVICE SOME APPEAL, BUT IT CAN MAKE IT MORE EXPENSIVE TO OPERATE. THE DESIGN HAS THE SAME CONSIDERATIONS AS THE ESCHER MACHINE EXCEPT MUST OPERATE UNDERWATER, INCREASING THE LIKELIHOOD OF ACCIDENT AND DISAPPOINTMENT. THE SPECIAL COMPLEX ANGULARITIES INVOLVED MAKE THIS DEVICE ONE OF THE MOST DIFFICULT TO EVER GET OPERATING, THOUGH IT IS BELIEVED TO WORK IF BUILT WITH HIGHLY UNUSUAL DEGREES OF PRECISION. EVEN THE INVENTOR IS NOT ALWAYS SURE HOW IT OPERATES, OTHER THAN BY COMPARING IT WITH WORKING EXAMPLES OF THE ESCHER MACHINE WHICH HAVE BEEN INVERTED, AND USING A BUOYANT BALL IN PLACE OF A WEIGHT. THE EXACT ENERGY OUTPUT IS UNDETERMINED BUT IT IS BELIEVED TO DEPEND ON THE DEGREE OF BUOYANCY OF THE BALL RELATIVE

TO ITS MASS, AND LIKE THE ESCHER MACHINE, IT MIGHT BE RUN USING A SERIES OF BALLS INSTEAD OF JUST ONE. ONCE THIS IS GOTTEN GOING, IT MIGHT HAVE A LOT OF POTENTIAL IN PRINCIPLE THOUGH ONLY IF THE PRECISION CONSTRUCTION STAYS IN PRECISELY THE SAME CONDITION. THIS IS NOT A GOOD DEVICE TO COMBINE WITH EARTHQUAKES OR POOR CONSTRUCTION.

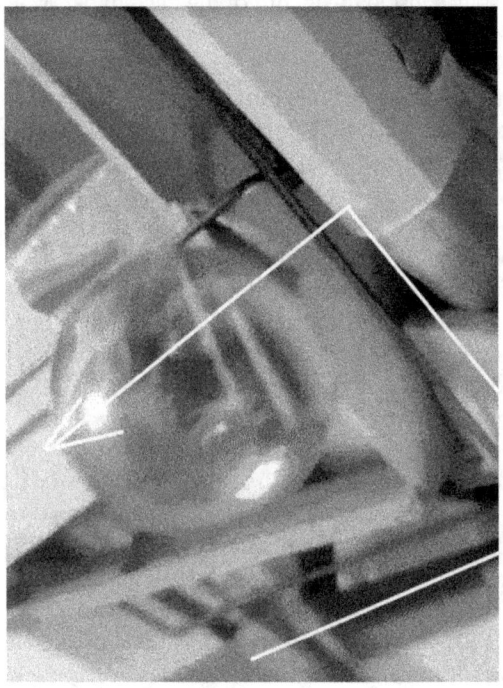

INSTRUCTIONS: BUILD AS ESCHER, ONLY UPSIDE-DOWN, USING A BUOY IN PLACE OF THE BALL.

Unnatural Torque (self-driving cars)

I have made a list of all the absolute requirements that must be followed. It turns out they are all required, it was a bit of a lucky experiment: (1) Supports for fulcrum are about 2X back, 1X front, angled at 45 degrees downwards from fulcrum, (2) Fulcrum is supported by a vertical pin attached to a horizontally 45-degree angled fixture which can rotate 360 degrees within the fulcrum unless inhibited by the lever, (3) 45-degree angled fixture is also supported at close distance inside the length of the fulcrum bar by a fixed hole running parallel to the lower end hole on the fulcrum support bar. (4) The fulcrum support bar runs between the lower and upper ends beneath the 45-deg angled fixture and is NOT firmly connected to the fulcrum pin AND NOT firmly connected to the pin rotating sideways and supporting the fulcrum pin, the lever also rotates laterally at a specific angle due to the angle of the entire apparatus, (5) The upper end of the fulcrum supports is supported by over 0.5 inches maybe significantly more of additional height underneath, various angles should be tested to maximize utility, (6) The back end on the lower side is supported as shown by a different 0.25 in approx. additional height which may be very shallow, but seems to help, (7) Short end of lever is heavier, (8) Long end has small amount of weight, (9) Weight on short end is attached ABOVE the short end of the lever, and can be turned 45 degrees vertically towards the lower end to create motion, (10) Longer end is supported by very smooth bar at significantly lower altitude than the fulcrum, (11) The smooth bar is kept some distance inside the length of the long end and the lever itself is smooth and straight, (12) The smooth support bar is kept at a sufficiently shallow angle as to allow motion to take place. (13) It may help to rotate the entire device slightly counterclockwise from above to achieve the ideal lever position.

INVENTOR'S ONLINE JOURNAL

July 10, 2018. Some of these concern competitors. Not all are my own inventions, but I would say most, at least originally.

UPDATE JUNE 25, 2021: Perpetual motion is probably too public to require a secret, and scientists continue to act very skeptical, though it has been months or years since someone with a real science background has given a second though to my work in any reasonable way.

(Main links: Perpetual Motion)

More Extensive Records: **The History of Perpetual Motion Machines**

A Few Notes on Correspondence:

"There could be more research definitely from a science-based point-of-view. Even the smallest efforts like blog posts, mentions in self-published books, and online discussions can make a difference. One of my disappointments is there has not been a strong presence for my work on overunity.com" ---Message to Ramiro A. S. La Rotta (Academia)

Bulletin

May 12, 2020:

Counterweight lifts lever along slight upward incline, uses momentum to activate ratchet which sets lever slightly above track, lever then returns by its own mass, ratchet is released. Lever and any attached ratchet parts should have a mass of 1 plus additional structural mass of 1, with the counterweight being equal to the max leverage / 2 + 1 to min leverage + 1 and positioned on the short end. Long-end leverage is expressed in units of the entire short end, measured to the midpoint of the counterweight. (May 12, 2020).

May 9, 2020 (Or, it's May):

Improved variation on Counterweighted Magnet design: a strip of magnets pulls a ball upward slightly along a wedge using >0.5 to <1X force. The ball is then able to drop downwards using its slight advantage of 1X force.

Jan 24, 2020:

Another idea is to use horseshoe tubes with weights that float above where the opposite water is. The net height permits motion because the opposite buoys are sent upwards along the curve of the tube. A variation on this is hollow helmet- or pole-shaped buoys that allow extreme gain on buoyancy. A promising idea might be to use thin drooping chains with buoys, perhaps sealed in a plastic tube, if poles don't work.

---Nathan Coppedge, comment at: VE Project: Unbalanced Spokes Lever. Perpetual Motion Machine

Note on 1/2 m*d experiment, 2019-10-15:

(It is the same as if you would pull it horizontally with the same force?)

Sort of, that is the motion is close to horizontal if that is what you mean. But, if that is not what you mean then, no, several factors: 1. The LARGER stick moves upwards, 2. If what you mean is $f = ma$, then no kinetic acceleration takes place as the weights begin close to rest, using only their mass. So, those factors may be something to notice. Since the string is taut at the beginning, not much kinetic force will be translated, so $f = ma$ is likely the wrong equation. It is closer to leverage using effective mass or a neutral pulley, or weights in a balance. However, we know neutral pulleys do not move from rest, and balances with equal masses always end in equilibrium. Better than a balance tends to mean perpetual motion.

---Nathan Coppedge, Message to Overunitydotcom [Youtube channel]

May 28, 2019

The estimate from the article of about 1.1 degrees is in the ballpark. I was just writing about applied perpetual motion, maybe more should join in the fun. Great minds think alike... Google was working on cold fusion recently...

---What is your theory behind Graphene's "magic angle"?

APRIL 26, 2019

Perhaps saying what many intend to say,

"I not any understanding my brain broken broken.."

--Nix Da, comment at Vertical Perpetual Motion

...

"Oh thank you for the very much information. Right now were designing a blue print on how it could be receptive to come about. Were using the college lab to use the 3D printer so we can see what we could design. The main goal right now is just to come up with a prototype for the project and then we can see the kinks and errors that exist so we can fix it" ---Judah Yodhh, May 25, 2019

STATEMENT

I am currently working with Judah Yodhh, who claims he has found someone who may be willing to create a better version of my experiments. My experiments, for their part, are very promising.

For once in history thus far, the machines appear to be completely realizable.

[Response to General Criticism:]

"Set it on loop. I just realized the place that holds the marble is lower than the edge of the grey member, therefore it probably works, as the lip is not required!" —On the Vertical Lever Video (2019/03/25)

"Re: 'The core problem' once it has natural two-directional motion from rest as [shown], it is only a matter of proportionality, my gentleman."

---Message to Matthew Egan, March 15, 2019

You don't like the way I calculate... However, if there is a significant imbalance of weight accounting for effective mass and effective leverage and the angular modifier, objects in a balance will seek to equilibrize.... And this is exactly what I've observed.

NOTE AND UPDATE ON SWIVEL LEVER: The preferred angle of the Swivel Lever has been found to be a parallel angle with the track or with a very slightly more upward angle and an angled basket as shown in videos. —2018-12-15

STATEMENT DECEMBER 3, 2018

My K-nex constructor toy sets cost less than $90 all together when they were first bought, and resulted in the Invention of perpetual motion on Oct 11, 2018. However, I have found the pieces are too flimsy for building a model that is worth more than a proof.

---What is a <$100 purchase that has paid for itself many times over?

...

Now you can see why the machines cannot be theorized with ideal physics, only accounting for the mass of the lever itself with the mass of the lever (as opposed to the marble) treated as constant rather than compounded with leverage.

---Message to Simon Derricutt

Perpetual Motion of the 1st Kind

10/2018: Shortlist of the moment: Swivel Device, 1st Fully Provable, Vertical Lever, NIBW4, Crescent Lever, Escher Lever, Escher Delta, Escher Machine, Modified NIBW6, Modular NIBW 1 & 5, Double-Disk Device, Coquette with Fixed Sidetrack, Double MMM, Vertical Slant Device, Escher NIBW3

AWAITING INPUT FROM MODEL-BUILDERS AND 3-D PRINTERS! (AND ENGINEERS AND PHYSICISTS!)

Precautions for 3-d Printers of Perpetual Motion

3-D PRINTING PERPETUAL MOTION, PART 2

"If it works I'll think of it as your design too since it is based on your discovery"—Jer Ram

2018/04/02, Possible new collaborator named Judah Yodhh says he knows of a possible future builder.

NOTE, 2018/10/07: I seem to be the only one working on it, other than Jer, AB Hammer, Reidar Finsrud, and at one time the V-Gate guy. Perhaps I can earn credit somehow if I am involved with the first working model, assuming Finsrud's is a hoax (Finsrud's seems to rely on purely horizontal motion).

NATHAN COPPEDGE'S PRACTICAL PERPETUAL MOTION HANDBOOK

UPDATE 10/2018: STILL HOPING FOR SOME KIND OF CREDIT.

STATEMENT 2018/09:

There is some good evidence, and the most basic principles seem finally discovered. It involves overlapping multiple fundamental efficiencies so as to create motion and overcome friction. I welcome assistance from volunteers ready to make videos verifying my successful or partially successful experiments:

- 2018/10/29, IMPORTANT STATISTICS ON ALL OF THE MOST PROMISING DESIGNS PUBLISHED ON QUORA **HERE**
- 2018/09/18, POSSIBLE NEW COLLABORATOR, RÉMY D.

Actions by Jer Ram:

- ATTEMPT TO MODEL PROMISING CRESCENT LEVER DESIGN: **Are we ready for the rest?**
- ATTEMPT TO MODEL PROMISING VERTICAL LEVER DESIGN: **Track in progress still mostly incomplete**
- JER RAM CONFIRMS NATHAN'S NOV 2013 SUCCESSFUL OVER-UNITY

EXPERIMENT 1, Nathan Coppedge's discovery

...

More minor updates:

(2018/11/13): Coppedge begins requesting parts from John Mills, a 3-d printer (Later note: Coppedge was unsatisfied with the initial quality of the service and decided success through this 3-d printer was too much of a longshot).

(2018/10/06): Jer Ram shows he is capable of printing some precision parts. (Later note: Jer Ram was not able to print big enough parts, and his primary commitment was 3-d printing, so we have stopped collaborating)

(2018/08/01): Currently encountering technical issues with the 3-d printer.

Journal of Research:

(Descending order of initial research)

JOURNAL FIRST SEMI-AUTHENTIC PERPETUAL MOTION COMPUTER MODEL

JULY 10, 2018

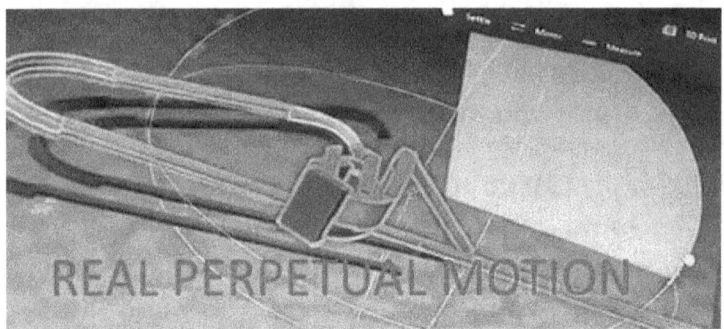

ABOVE: Image from Real Perpetual Motion Group on Facebook. *Note: This particular design was found not to work, but it seems often particular difficult cases are hard to understand without experiment.*

I'm hoping Jer Ram has the answer.

He is my 3-d printer for what may be the world's first perpetual motion machine.

However, it is in the very early preliminary stages. Do not bother him for a few weeks.

JULY 12, 2018:

PERPETUAL MOTION MAY BE 3-D PRINTED SOON

Basically, very close to the exact ratios must be kept for it to work, but the masses are flexible when we keep the ratios. Therefore the counterweight is not

NATHAN COPPEDGE'S PRACTICAL PERPETUAL MOTION HANDBOOK

really flexible by itself, and therefore the leverage ratio must be closely maintained. As a result the mass RATIO is not flexible, and as a result the lever structure unweighted must be kept ulta ultra lightweight. These conditions essentially must be met.

You see, it is an exceptional case. That is the hard line.

---Nathan Coppedge

If we can build this probably even black cats will look lucky, but that doesn't make it impossible. ---Nathan Coppedge

JOURNAL VERTICAL LEVER:

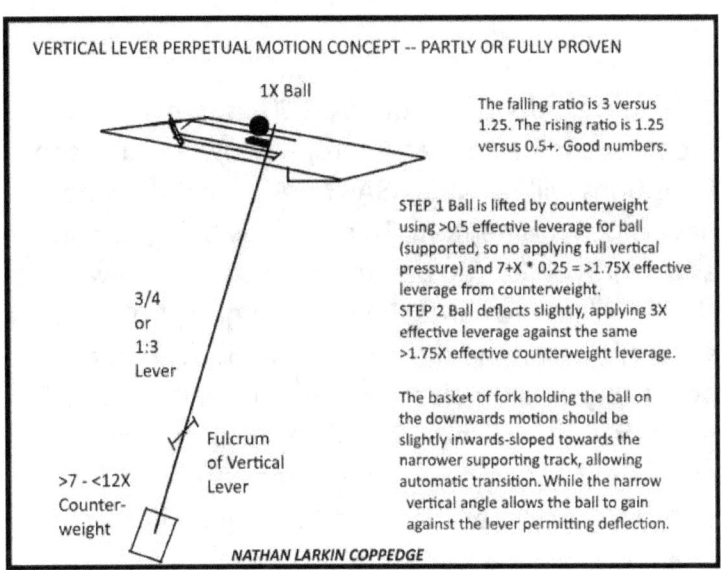

JULY 28, 2018:

I guess if you could modify the cone-shape [basket] to be more cubical off to the side, but downwards-sloped towards the narrow track, cut off most of the inside of the cone, and introduce a wall that the marble will glide against, and a cylinder attached to the base of the currently-cone-shaped thing that will snuggly receive a k'nex member, and lower the operating altitude of the cone-shape so there is much more room and the angle can deposit the marble in the beginning of the track, then that is what we need.

Comment @ Track in progress still mostly incomplete

....

1. The lever *does not rotate*. 2. Thus, the inside of the track can be connected from below, 3. Ultra-short transitions will be NECESSARY, because it does not have much momentum. Think of it as being as narrow as a sandwich, 4. The basket will need a downwards slope directed towards the beginning, so it will need a sufficient wall to hold the marble in. Think of it as being very small, with a complex shape. It just needs to prevent the marble from falling and deposit the marble in the beginning of the track. There really sort of *isn't* a track for the downwards drop, just walls that I am calling a 'slot'.

---YouTube Message from Nathan Coppedge to Jer Ram.

...

New detail with Vertical Lever: using a slightly protruding basket that distances the ball when falling from the lever can help increase the height of the marble at the base and lowness of the basket at the height, while maintaining appropriate lever and track ratios. However, care should be taken to make sure the basket does not gain height as it is free-falling, as this may prevent it from moving.

You may have to use a somewhat large marble... This is reflected in a new diagram for the Vertical Lever in which the track is somewhat sharper than before and the motion of the lever is shown to be proven to have zero loss, and possible gain with even steeper track or with longer leverage ratios. Note the track is still significantly less than 22.5 degrees to allow the support effect.

---YouTube Messages to Jer Ram 2018/08/03

...

What I meant to say before was that the marble should be a little big so that the point in the basket the marble rests on does not lose as much altitude as the stick on the return, but the marble can still be pushed by the stick at the base. This may be important.

---YouTube message to Jer Ram 2018/08/04

Note on the Perpetual Motion Proof of Concept (Partial) featuring the Vertical Lever:

[The recent] experiment was very difficult and required many, many very slight modifications in the direction of the lever, exact configuration of wire, altitude of the grey support member and the perpendicular grey push member, and some very careful consideration went into the shape of the cutout platform and the height and sturdiness of the platform as well as the length and tensile strength and lightness of the lever, and of course the exact weight of the marble matters relative to the lever and counterweight. Also, friction and stickiness in the marble, wire, and track must be avoided or there are dire results.

---Nathan Coppedge, YouTube Comment, 2019/03/21

...

JOURNAL CRESCENT LEVER:

Project may be switched to the Crescent Lever due to the size constraints. Simple parts may be printable larger than 3 inches.

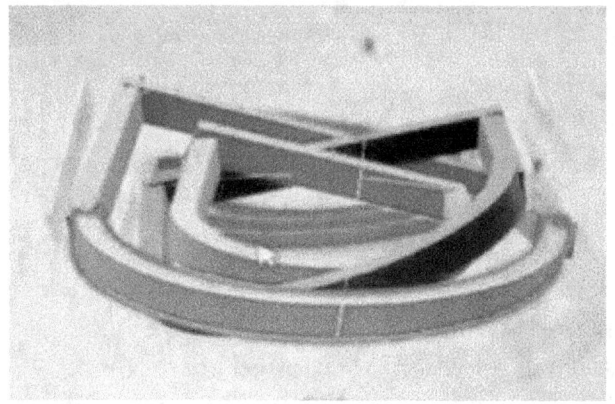

ABOVE: A model of the Crescent Lever crescents. Credit: Jer Ram, YouTube.

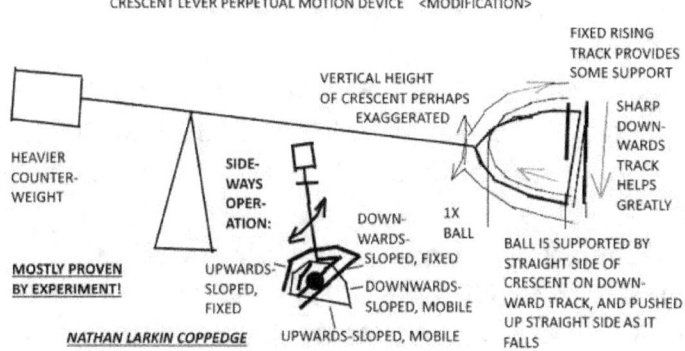

The neat thing is **the height of the vertical drop can vary**, so the horizontal loop can be as long as you want. *That's* the secret that I am reluctant to reveal even to you, until now. *The verticality is completely flexible.*

---YouTube Message to Jer Ram 2018/08/08:

...

You're headed in a good direction, but here are some modifications I might need: 1. I'm accustomed to using clockwise motion for the crescent lever, I'm not sure if the counterclockwise motion was intentional.
2. **The lever attachment should always be upwards-directed when positioned correctly on the straight side. I know this is counter-intuitive**, but **all the motion that occurs on the straight edge should occur due to the marble's mass and the angle of the track PUSHING THE LEVER *DOWN* AND *LIFTING* THE COUNTERWEIGHT**.
3. **Consequently, we will want a 22.5 degree or greater *DOWNWARDS* slope for the track *GOING THE SAME WAY THE LEVER IS GOING UP* *[on the straight side]*.** This covers motion along the straight side. The 12.5 degrees slope for the attachment is fine, but **the crescent attachment should be *FLAT AND TILTED* WITH ABOUT A 12.5 DEGREE DOWNWARDS SLOPE THROUGH THE CURVE while the track is *upwards* (BOTH ARE REVERSED BETWEEN THE STRAIGHT SIDES AND THE ARCS OF THE CRESCENTS).** Once again, **this is the only way to create UPWARDS MOTION....** 4. The exact angle of the upward track section is not as big a concern, however it is maximized when it is slightly less of a pure semicircle, maybe more of a

horseshoe shape so that it gains more altitude relative to it's angle. **I would prefer clockwise motion, but if you say counterclockwise I will try to work with that. It should work either way or it doesn't work...** The horseshoe shape should not be very long because we don't want a really long lever, but a little bit more length in the curve would be good. Also, 5. I know we don't have much room, but if you can make the width of the crescent almost as narrow as possible while maintaining something similar to a horseshoe-semicircle in the attachment, that would be ideal. Keep in mind that although the attachment will be attached to the lever below somehow not directly at the level of the track, **the lever itself will probably be horizontsl with adjustable height.**

Also let me know if you find a way similar to my earlier suggestion to connect the inside and outside of the track from below the straight side, some holes or tubes for k'nex should be fine. I was also not sure if the **semicircle... elongated slightly, I think I mean just a little bit more, it can be noticeably the longer dimension so as to maximize the angle of the vertical drop relative to the upward slope of the crescent track. / The crescent track can be shallowly upwards-sloped except we *NEED* a sharp downwards slope on the *straight side of the track*.**

---YouTube message to Jer Ram, 2018/08/11

...

Looking good! I expect these angles are workable. **There should be more smoothness ultimately in the transition to the upper track,** otherwise we should get to all the other stuff basically, except that I would prefer cylinder holes for the outside of the track, rather than clip tabs if possible.

Previously I was being very long-winded (I have deleted a few comments), but basically you've made marvelous progress, and this may be the world's first functional perpetual motion machine when we finally have it constructed!

A subtle thing I have figured out how to do is to make the end of the upper crescent slightly downwards-sloped when the crescent lever attachment is still downwards-sloped as well. We don't have to apply that subtlety yet for a crude model, but it may be worth trying that design if the model somehow doesn' t work. But very likely if it doesn't work it's just something wrong with the lever or weight ratios.

Very good work, and I hope my depression isn't so depressed as to kill this thing.

NATHAN COPPEDGE'S PRACTICAL PERPETUAL MOTION HANDBOOK

—YouTube message to Jer Ram, 2018/08/19

...

Sorry, I meant to say that the holes should be vertical, that is, coming from underneath but not fully penetrating the track. This might require cylinder protrusions. This is what I'd try at first Sideways holes might result in obstruction and create more difficulty with k'nex due to crescent. For the outer crescent the holes should be at about 1/3 positions. I intend to attach them by vertical bars connected by a atraight member underneath. The inside crescent should also have two holes, this time at the far corners of the crescemt, hopefully sturdy enough to hold the crescent from there. If you think it will be brittle, or un- sturdy, **you can add a single hole underneath the middle of the inside track which because it is the inside will not interfere with the lever.** I will find some way to connect that, you may as well make that hole if you can just for added possibilities. Additionally, the corners of the larger and smaller flat side should have holes which will allow the inside to be attached to the outside, yet accommodating the smoothness of the 22.5-degree downward slope. I am no longer sure walls will be good due to the need for flexibility in the ball. **We will try the single vertical tube for the crescent lever attachment and hopefully.that will work for test purposes.** My hope is this arrangement works better than the slanted lever. Also, it's okay if the holes are a bit loose as long as the track doesn't jump out, **some of the structure will be provided by the k'nex [constructor pieces]. Another factor is that the cylinder holes should not create bumps on the inside of where the track is being used, as the mobile weight in some cases may hang down** significantly. Great work overall, I like what you are doing…!

---YouTube message to Jer Ram, 2018/08/28

…

The walls look very good. Even if the walls don't print, **we can try it without the walls as well, as I think the track is wide enough to provide continued support for the ball. However, we should try to print with the walls if we can, in part because it can improve people's psychological response if the device is larger or has a larger number of necessary parts. It improves people's ideas about how sturdy and reliable the machine is, and improves the likelihood that people do not think it was impossible that it was not built** before.

I wish you luck on the next design & construction steps. I still think you're better with materials and construction than I am. When I made some of the measurements I was just referring to online tables made by the ball bearings manufacturer which are quickly found with a Google search. I have had a few years to practice weight estimations since considering my Buoys device from back in 2005. The information on aluminium buoys was very technical because it's expensive stuff and it's in high demand with professional oceanic divers. Pretty funny to think about. Anyway, I am not always the best at equations but I try to cover all the data that helps when it is most needed, rarely at other times.

I know perpetual motion deserves the most professional treatment, and if I had a lot more money I'd be paying professional engineers and designers to back up my evidence. But, unfortunately that's not an option so it's very lucky that I have [Jer Ram] to work with, although frankly I think of you as a kind of engineer with my situation in mind.

---Youtube message to Jer Ram, 2018/09/08

...

Thanks for the update. You can just post a shorter video with the printed object, no need to keep posting videos on print bugs. I suspect anyone else trying to model this will have a different problem if they have a problem at all, and in most cases they will be using different printers / and or software. You have already shown that the printer sometimes prints complicated shapes. I know it has (or you say it now has) a different bug than it had before. But I am really just interested in two or three things, directly to the point: 1. Your ability to print physical objects, 2. Your ability to model specific physical objects, and 3. Your ability to model physical objects in general. If you just show me that you can print an object of appropriate scale that is moderately complex, that finishes steps 1 and 3. I realize that's what you're trying to do, but that is basically a hardware problem which has no relevance to my side of the process. I am trying to be supportive just because you're spending your money, and I am very interested in the end result, but basically I am just interested in your ability to print, and creating usable end-results. I have very little technical interest in how you get the printing done, I am just trying to be supportive so we get real results in the end-product, kind of like collaborating on an academic paper, or a business relationship---very hands-off. I do not have any *real* experience with 3-d printing is the scary fact, and I do not ever plan to buy a 3-d printer. For the rest of my life I plan to always rely on others if I do any 3-d printing, and you are the person I am able to work with right now. I wish I had an academic collaboration or a corporate sponsor, but I don't. So, I am very glad I have you to work with, and I realize technical problems are a big concern, but if our process works at all these technical problems should not last forever. You probably have more intuitions on how your 3-d printer operates. I don't know if you know whether we will get an accurate end-result. What I know is, if you can print the physical object, that is the same thing as printing what I want you to print, and if there is a problem with that other than it's structure and precision, then it's something wrong with my design. The design is not on your shoulders, but I am hoping you have the resources, mental and physical to make the physical product workable. 90 mm might work, 95 is better. In some cases we may need to make all the track angles more shallow in that case to assure there is sufficient variation in real units between the straight side and the horseshoe. Although real-unit variation is relative, it can have an impact on the constant of friction between larger and smaller scales---friction is a bigger factor the smaller we go. Therefore, we should test printing at the largest scale we can get, or pretty close, that will allow smooth motion, even if we can't print proper supports. Any kind of improvement of evidence would be invaluable at this point. I would prefer we at least duplicate my previous Crescent Lever experimental evidence with 3-d printed materials in the next year, even much sooner if possible. If we can duplicate the experiment we may be able to cross the line into real perpetual motion. Some of the barriers at this point I understand are mental, social, psychological, physical. However, with proper design and construction the only barrier is creating the physical model, and that is something a 3-d printer is designed to do---the beauty of 3-d printers is they should be able to print anything physical of a certain scale. So, logically we seem to have all the cards necessary to improve on my experiment, and if that is the case, that is not so different from perpetual motion, and perhaps we will be able to cross that invisible barrier that divides mortal and immortal civilizations. It looks psychologically challenging, but when we break it down into workable steps, we can divine that all that remains is physical construction and a few tweaks. It would be great if I had the best companies and the best engineers working on my mission, but right now if there is anyone like that that is fully cognizant of my work they are not even on my side. So, I have tried to minimize physical stress on anyone else including you and any interested engineers and corporations, but we have to realize, all that remains is physical construction and a few tweaks. The experiment is done. Many experiments have proven the impossible. But the world is lazy, and the professionals are not even aware. So, I am hoping you, perceptive and insightful as you are, are able to pull through, or else I will need to look in higher places or over a longer time, and it is painful to think that a project that is already a complete success could end in abject failure, simply because the majority of people we trust could be subjected to a deep, abiding, and penetrating

Socratic critique ending in the conclusion that Einstein was not so bright, and von Neumann was not so bright, and Tesla was not so bright, and DaVinci was not so bright, simply because they did not invent perpetual motion. However, if perpetual motion succeeds, a certain kind of existential foolish genius will certainly be inevitable. And that terrifying lemma, for all those that have reveled in humanity's destruction, that all of humanity's practical problems will ultimately be solved.

---YouTube message to Jer Ram, 2018/09/29

...

Oh, I see, we probably need [attachment points] for connecting the inside and the outside [of the vaguely protractor-shaped track].

---Youtube Message to Jer Ram

...

I know these may just be tests, I am impressed with the quality but hoping there will be a flat side ultimately. If this works great, maybe you should test it if it is your design however.. What I suspect--I know this sounds like nagging---is that there will not be a sufficient downwards slant on the falling side to create two-directional motion if it is round. Basically you didn't try a slant Coquette or slant NIBW6 yet so

you don't know this lesson. I am actually nominally impressed with your learning ability as we're basically on the same pages just a few years apart. Anyway, it's not a total failure to print the circular design,
but **[Jer's adventurous circular version of the Crescent Lever] does not pass one of the last criteria, which is motion from rest at every point.**

---Youtube Message to Jer Ram 2018/10/07

...

For the outer and inner track additional structure is okay to prevent wobbling. It is only the (crescent) lever attachment that has to be not very thick. The crescent attachment could be a bit thick, it just shouldn't be extremely tall like (<) 4cm about. That may require some testing. Just think about it similarly to overlapping triangles--how thick would a triangle have to be to stop moving in the middle of other triangles? You'll probably get as much as 2X height out of every gap distance, so a 10mm.gap on one side would produce a workable height of about 16 - 18 mm maybe, or 10 or 12 to be safe. I know this could create printing problems, but just so you know it could work thicker it just has to not be extremely tall in the lever attachment, and the tube or such for the round side of the attachment needs to be secure enough that the lever can attach from.far below, so the thickness of the track elements is less of an issue that way--levers are fairly extensible if built properly. As hard as possible plastic is good, but if it is expensive save it for improved models. Keep in mind I think you know a crescent (not circular design with a flat side will be necessary, it's a bit hard to explain why but suffice to say there is a critical difference involving advantage of mass against angularity (unbalanced asymmetry). The flat side 'helps' but it turns out the help is necessary. It is one of the hurdles to realize this. Anyway, I'm glad to see you are making progress. We can try cutting the plastic and taping it if we have to. What I meant to say is the straight side produces a much sharper angle for the altitude, or we are forced to use a scoop shape or lose altitude relative to the beginning of the crescent. The amount of drop is maximized with a straight side. Using a slightly narrower horseshoe increases the advantage even more. In fact, we could almost use a descending fork lever, except this makes the inner turn difficult.

---YouTube message to Jer Ram, 2018/10/25

...

JOURNAL SLANTED REPEAT LEVER:

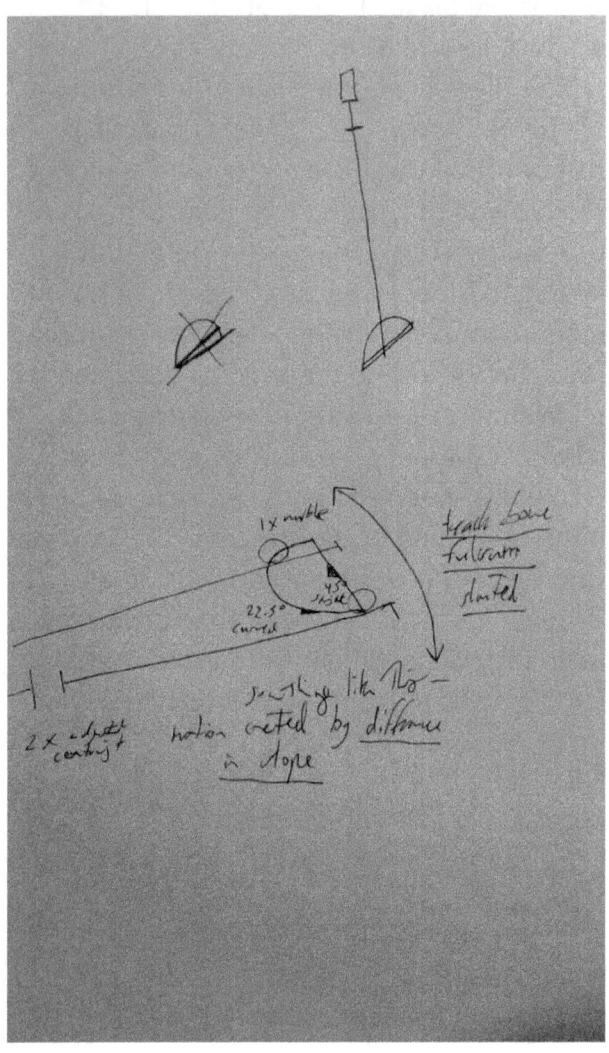

Earlier outdated version:

Some signs of success with a slanted version of the Repeat Lever. This is something we could try that could be made very small. The upward part of the triangle must be downward sloped at less of an angle than the upper end of the lever and a cut-off elevation is used to permit the marble to have support at every point. This could be very easy to build and test. I'm not sure of the lever angle or exact weight ratios yet but something probably works.

---YouTube message to Jer Ram 2018/08/10

...

Improved, updated version [pictured]:

New Secret of the Slanted Lever --Use an arrangement similar to the Crescent Lever, with the fulcrum lower than the entire track to achieve horizontal advantage, but on the horizontal more towards the midpoint. The horizontal advantage will create the upwards motion with sufficient counterweight mass (2X+), while the return will be less of a concern due to the 45 degree angle of the straight side of the track. The curved upward slope will only be about 22.5 degrees accounting for the curve, due to the longer distance. Thus, there is a justification of motion in measurements of 2X or greater counterweight mass with >2X leverage, creating ratios such as (in the case of 3:1 leverage and 3X counterweight mass somewhat conservatively assuming very lightweight lever and 0.75 effective mass for the ball due to the properties of wheels) --> 2.25X ball rising : 3X counterweight : 3X ball falling. Accounting for the weight of the lever, however, Rising effective mass might be 2.5, and falling might be 3.25, yielding a workable ratio due to the properties of a balance. Alternately the straight side might be as shallow as 22.5 degrees and the curved track even shallower.

---Facebook comment, 2018/08/14

...

JOURNAL REVERSED TRIANGLE REPEAT LEVER 2:

The best ratios I have found for this are:

4.5X counterweight (1X distance from fulcrum).

1X marble 2.5 - 4X distance from fulcrum.

1X maximum long-end unweighted leverage (with no counterweight or marble, e.g. mass of about 0.25X).

This yields: 5X matble resistance unsupported.

4.5X counterweight resistance (constant).

3.62X marble resistance supported using estimated 0.75X mass on upward incline.

'SECRET': Now imagine that the lower track can begin below the fulcrum and the upper track can be nearly flat. This may pose extra special advantages.

...

After recent investigations I see that you seem to have been right about an option of reversing the triangle in the repeat leverage. I have posted a drawing and some data which suggest something similar to a right triangle in which the 90-degree angle is in the upper inside corner may be highly workable. This is another design that might work small and simple.

---YouTube message to Jer Ram, 2018/08/14

...

JOURNAL GENERAL FORMULA:

Expressed in ratios, from a minimum of compensating for max leverage * >0.5 effective mass + effective long-end lever mass unweighted to a maximum of max leverage + effective long-end lever mass unweighted = Approximate exact range of working counterweight mass in all devices. For example, for a 2:1 lever the counterweight will be about 1X mass + lift +

weight of lever, whereas in a 3:1 ratio the counterweight will be about 1.5X mass + lift + weight of lever. These values will yield ranges of >1.5 rising ball : 2 counterweight : >2 falling ball and >2 rising ball : 3 counterweight : >3 falling ball respectively. Thus increasing leverage by 50% results in almost 100% increase in the allowable mass ratio for the lever when it is unweighted on either end, creating an efficiency. [later revised, likely some unusual variations use different ranges such as 1 : 0.5 as described in Perpetual Motion Statistics].

---The Equations by Coppedge

...

JOURNAL FIGHTING DEPRESSION:

My mood has improved significantly recently. I think the slightly longer loop in the crescent and slant lever can make a big difference with low friction. I'm not sure I expect success immediately, but at the same time I do not think it is impossible to just pull it off in an odd moment. There is not nearly as much of a differene between dysfunction and a functional perpetual motion machine as many think. There's no need to view it coherently, it is just a matter of whether the chain reaction happens to work, and I have been very close before. The last increment of cheating may be the most difficult but that doesn't mean it is impossible. I have added layers of cheating physics before, and while it may depend on the design that is not to say the design is inflexible. And many of them do work to demonstrate simple over-unity if not perpetual motion yet, so logically it is possible. How possible or 'easy' it is is still a matter of debate, and it does not help that many educated people assume the most basic efficiencies are impossible even though I have proven them again and again since Nov 2013. In any case, I am vacillating between expecting success and expecting failure but I think the track record shows someone will eventually do it unless the laws of Newtonian physics are utterly dispensed with. So, it is a bit of a catch-22. What proves perpetual motion is utterly Newtonian, but is interpreted by scientists as being the very antithesis of Newton. However, physical laws and perpetual motion machines should not be hampered by mere scientific thinking and the habit for confirmation biases in new endeavors. There is evidence, so in a simple categorical sense it is certainly possible. So, I think that is why I don't feel depressed.

---YouTube message to Jer Ram, 2018/08/18

...

JOURNAL SPIRAL (SPIRAL CONE) DEVICE:

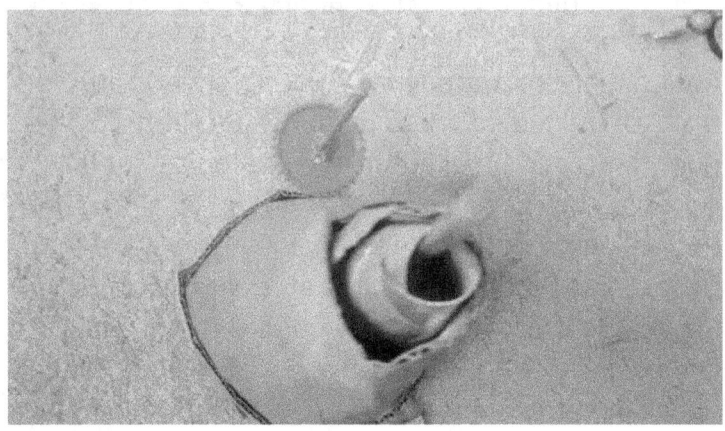

Yes, it might be worth trying something like this with the crescent lever track, it might work in a similar way.

---YouTube message to Jer Ram 2018/08/30

JOURNAL ESCHER DELTA:

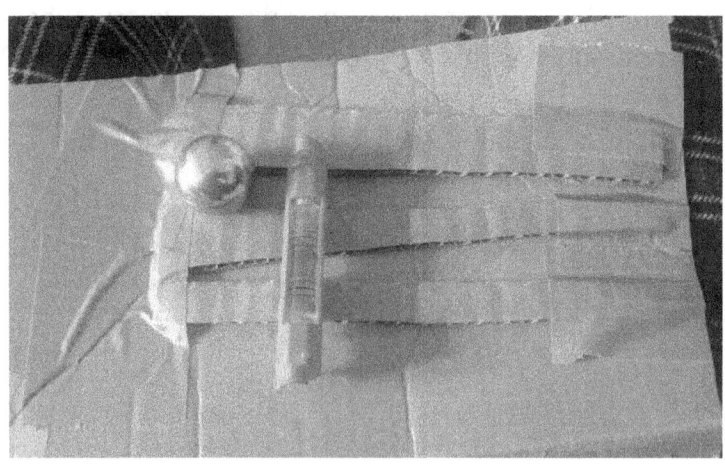

NATHAN COPPEDGE'S PRACTICAL PERPETUAL MOTION HANDBOOK

Promising "Escher Delta" has shown 2-directional motion when the wide ends are bent upwards , a wider track is given altitude, and preference is given to a direction of motion.

---Nathan Coppedge (Facebook)

...

JOURNAL SLANTED PULLEY WHEEL:

Support is used as a cheating method. 360-degree motion is permitted by pulley. The ratios are 1 : >0.5 -

<1X with the counterweight being lighter using the 1/2 mass* distance rule on the round weight during the slanted upward motion. Thankfully the angle can be modified so the resistance is 100% flexible.

Real Perpetual Motion (Nathan Coppedge, Facebook)

JOURNAL MATERIALS SCIENCE OF PERPETUAL MOTION MACHINES:

Dear Dad,

Every material has some disadvantage.

Metal is hard to work with or bends.

Plastic is brittle or flimsy.

Wood is heavy or breaks.

Glass is heavy or brittle.

Clay and ceramic is heavy AND breaks.

Cloth and textiles are too soft.

Some of the best materials are really high-end and not really worth it.

Basically with perpetual motion some of the best materials are available, but are hard to work with.

It's a problem of making something look exactly how one designs it with the right specifications.

Essentially, half of it is design and the other half is construction, and there are some design-construction and construction-design issues.

I have done very well with my experiments by the common standard: showing cheating principles, but builders rarely understand how lightweight the structure should be.

Up to a certain scale plastic and thin hollow metal are essentially the only options for the structure, since it needs to be long, with tensile strength, not bend at all, and be as light as possible within those conditions. For example, heavy plastic is much less workable than light.

So, materials are pretty easy, but you have to know what to use and how to build properly with exactly the right materials.

So, 3-d printing and welding don't always work, and there aren't many other options.

Even a skilled designer with unlimited resourcefulness needs to know exactly how it works.

---Nathan, e-mail, 2018-09-12

JOURNAL NOT IF BUT WHEN 6:

NOT-IF-BUT-WHEN MACHINE #6

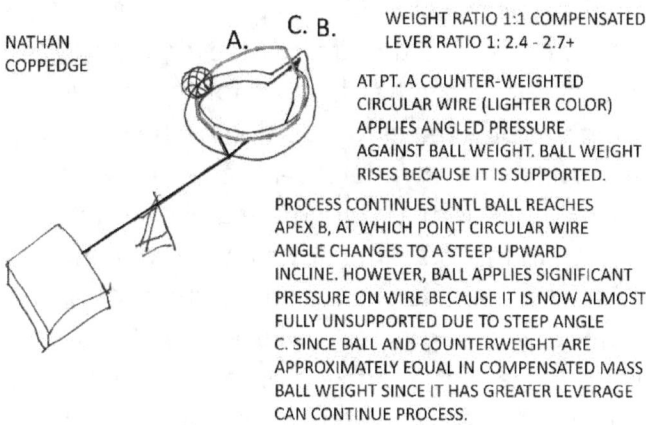

NATHAN COPPEDGE

A. C. B.

WEIGHT RATIO 1:1 COMPENSATED
LEVER RATIO 1: 2.4 - 2.7+

AT PT. A COUNTER-WEIGHTED CIRCULAR WIRE (LIGHTER COLOR) APPLIES ANGLED PRESSURE AGAINST BALL WEIGHT. BALL WEIGHT RISES BECAUSE IT IS SUPPORTED.

PROCESS CONTINUES UNTL BALL REACHES APEX B, AT WHICH POINT CIRCULAR WIRE ANGLE CHANGES TO A STEEP UPWARD INCLINE. HOWEVER, BALL APPLIES SIGNIFICANT PRESSURE ON WIRE BECAUSE IT IS NOW ALMOST FULLY UNSUPPORTED DUE TO STEEP ANGLE C. SINCE BALL AND COUNTERWEIGHT ARE APPROXIMATELY EQUAL IN COMPENSATED MASS BALL WEIGHT SINCE IT HAS GREATER LEVERAGE CAN CONTINUE PROCESS.

[Design is brought up and explained during difficulties with printing the Crescent Lever...]

I am reconsidering... and I see now that if the mobile portion is tilted in the same manner as before it mostly DOES have an advantage. So I take it back, this may be worth trying as a test, even if additional details may be necessary to make it fully working. Of course simplicity and parsimony is desirable, and it is admirable that you see that, although I don't want to simplify it so much that it doesn't work, but that has yet to be proven. In my experience with the Coquette

2 and NIBW6 it is a type of design that requires spiral or else skew. Skew may be workable in some cases, but I have found the best design with pure skew and no spiral is very likely the Crescent Lever with the flat side incorporated. However, there are major design similarities, as the NIBW6 is an earlier version of the same principle. Like the Crescent Lever, the NIBW6 incorporates a downwards slope for the mobile portion through the larger portion of the upward motion, and similarly there is a sharper upward angle on the mobile portion during the downward motion. However, in this case, the downward angle on the fixed track is kept very short due to the circular shape in order to create greater application on the sharp upward angle of the mobile portion. The application happens at the fullest extent of leverage and the steepness of the mobile spiral is kept to a minimum to allow the ball to have sufficient leverage accounting for the angle of the sharp upwards mobile portion. Modifying those insights, in some cases a 45-degree is used for the downward track to maximize weight application relative to the horizontal distance traveled, permitting a relative advantage on the amount of vertical distance elapsed on the vertical angle of the upward connector between the ends of the spiral. This design is very similar and could be worthwhile in this case if you find printing circles and spirals to be easier. It is not a very complex spiral for the mobile portion, just one loop and sort of shallow,

but I will need about the same amount of sturdiness as the previous crescent attachment. The design is about equally flexible, and I would not be surprised if it works, but I do not have as much experimental evidence so far on the NIBW6 due to the difficulty in building spiral attachments. I have made much more experimentation with the Coquette 2, once building it out of paper and sheet metal, but the Coquette 2 has been found to be unworkable unless a fixed track is incorporated to supplement the tilting structure. I would recommend avoiding the Coquette 2 for now, even though with the fixed track incorporated there may be a variation of it that is workable. The Coquette 2 tends to lead back to thoughts about the Trough Lever, RL2, and 1st Fully Provable, which may require large construction.

---Youtube comment to Jer Ram, 2018/09/16

...

Considering with perplexity the new proposed efficiency of this device of 200%, my conclusion was the efficiency happened simply because of low stress or what is called smooth motion relative to the amount of mass. ---Commentary, 2019/12/21

JOURNAL ABOUT THE LEMNISCATE:

Infinity is a somewhat good thing to attach to, but is not the whole picture. With perpetual motion infinity is a byproduct, not necessarily the most obvious feature. The loop of a perpetual motion machine is the obvious analogy to the symbol infinity. In many cases a circle or triangular form may be achieved instead of the lemniscate symbol. Think of it as an overlapping oroboros in which the second cycle loops with the first.

---Youtube message to Jer Ram, 2018/09/18

...

JOURNAL NOT-IF-BUT-WHEN 4:

"NOT-IF-BUT-WHEN" MACHINE #4

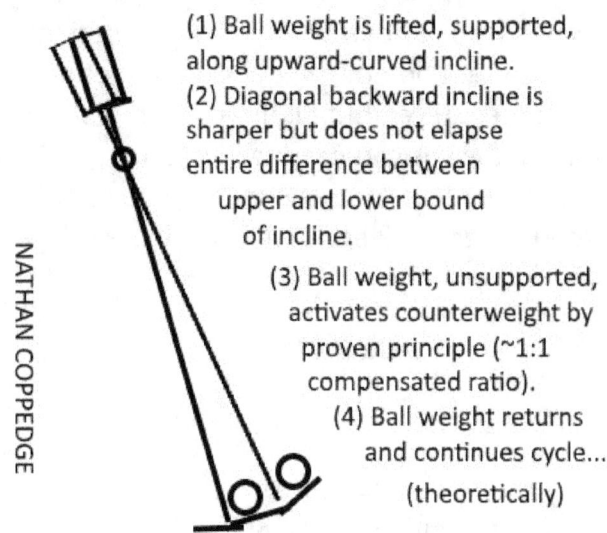

(1) Ball weight is lifted, supported, along upward-curved incline.
(2) Diagonal backward incline is sharper but does not elapse entire difference between upper and lower bound of incline.
(3) Ball weight, unsupported, activates counterweight by proven principle (~1:1 compensated ratio).
(4) Ball weight returns and continues cycle...
(theoretically)

Above: Diagram similar to experiments done in March 2016, both by Coppedge.

I heavily welcome any form of collaboration critical or otherwise. I hold that the above NIBW4 is a bit tricky but worth building. With proper ratio of weights and a slotted track it should be easy to prove the two-directional motion if you have a lever that doesn't flex while it is moving. Good luck, you're not necessarily gonna need it to prove just the two-directional motion. Remember the counterweight is a bit heavier and located on the opposite, shorter end of the lever opposite and higher from the ball/ marble in this case. Also remember the track here is upward-sloped away from the lever, with support for the marble coming only from the lever and not the track on the return. It is not fully perpetual yet, but I believe it can be. The experiment can be easily reproduced but has not been 'officially' verified yet by anyone. If you're doing a video, send me the link and I will include it on my channel and elsewhere.

---YouTube message to Rémy D

...

Dear Jer, I may have found a second person to reproduce experiments recently, Remy D. However, it is not clear he has any sophisticated equipment. However, he may have expressed interest in building a model next summer. Hard to tell if he was simply joking or if I caught the message late. Anyway, I have been hoping to have more collaborators however there have been very few openings. I'm hoping Remy D eventually reproduces the NIBW4 using a long lever, however I think if he does this he will be using crude supplies, I am not sure. Anyway, it is more reason to feel a bit hopeful as long as we're not under-the-weather.

---YouTube message to Jer Ram, *2018/09/19*

...

Thanks for the additional precision's. I will make a video of it (when it's done) be sure that i will share it with you.

---YouTube message from Rémy D

...

Experiments suggest if return has the same rate of slope or if altitude gain against length of lever from fulcrum has shown outward motion with capability of two-directional motion, the Vertical Lever Device and NIBW4 are capable of deflection or transition slopes and achieve perpetual motion. ---Nathan Coppedge, 2019-02-26

...

"JANUARY 28, 2021 Experiment with the Not-If-But-When 4 device suggests if we trust our senses, something similar to fully-working properties are possible. The operation involved in this case a short piece of plastic which was rubbing in an unknown way against a specifically arranged slanted cardboard. The arrangement is otherwise similar to previous models of the NIBW4 except that it has a lighter-weight structure than normal and larger, slightly heavier than normal ball on the long end. The motion seemed to be natural as it shows mechanical back-and-forth operation, however, immediate attempts to reach the same arrangement again once it was out of position failed. Thus, the arrangement seems to be very hard to reproduce or involve some kind of counter-intuitive choice. Otherwise, some type of environmental effect like wind or active neighbors might be the cause, although normally wind or neighbors would not cause mechanical motion in another room."---Nathan Coppedge

...

NATHAN COPPEDGE'S PRACTICAL PERPETUAL MOTION HANDBOOK

JOURNAL 1ST FULLY PROVABLE:

WHAT I CALL THE "1ST FULLY PROVABLE PERPETUAL MOTION MACHINE"
[MODIFIED FOR GREATER WORKABILITY] DIAGRAM WITH STATS BY FEB 14, 2021

9 degree angle is about 58.25 application. Max lvg is 2.24, Min is 1.73.
Min HvyMass = Max Lvg X 0.5825 + 1 = 2.3 mass units, (first check if able to lift ball to max leverage)
Max HvyMass = Min Lvg + 1 = 2.73 mass units, (check if still able to lift ball at min leverage)
Then there is a window of nearly 0.43 the mass of the marble to account for friction! In a balance! So, it works!

weight ratio: as one quarter, one penny 1 marble and 5 in of duct tape is to 1 marble
lever ratio 10.8 - 14 : 6.25
track angle ~0.5deg upwards-sloped
lever angle est 2 - 9 deg downwards-sloped.

NOTE: NEW LIP TO DIRECT FREE-FALL MAY BE IMPORTANT

NATHAN LARKIN COPPEDGE

B.1   A 1   B. 2   A 2

STEP 1: BALL HAS ALTITUDE TO APPLY PRESSURE TO FIRST MODULAR COUNTER-WEIGHTED LEVER. (B.1)

STEP 2: BALL RISES ALONG INCLINE, HELPED BY FIXED TRACK SUPPORT A1. UNTIL IT REACHES BEGINNING OF 2ND MODULAR LEVER. (B.2)

STEP 3: BALL WEIGHT HAS SUFFICIENT HEIGHT TO ACTIVATE LEVER AT SAME INITIAL HEIGHT, IN SPITE OF ITS LOWER BASE HEIGHT. PMM!

WORKING RATIOS in straight lever experiment.

I think I have used an adequate angle for the lever in each unit, unless it is just slightly too flat at the beginning of the units, but I think that could be handled by momentum due to the small difference.

---Quora Message to Jer Ram, 2018/09/30

...

I thought the units could chain react in a circle, particularly if they were set above the sidewalk on a higher level, then there could be outdoor energy generation on public land on every city block in some cities. That could be a significant source of power, since each unit can have a ball moving constantly in a big chain.

That's one of the reasons I give the 1st Fully Provable a rating of infinite infinity, because the number of balls in the chain is theoretically unlimited as long as you have more units to connect.

So, for example, it could run around the world and there would be no real way to count the number of balls, except in terms of their mass.

I can imagine the whole world becoming almost entirely urban by using energy for oxygenation, food production, manufacturing, and construction.

—Quora Message to Jer Ram, 2018/09/30

...

JOURNAL MINI-SCALE PERPETUAL MOTION:

It might be even more appealing [than nano] simply to make them miniature, like tiny cubes, bricks, squares, etc.

However, there is an energy paradigm that says that it is more efficient to build large using extreme mass, as this maximizes the amount of mass per mechanical area.

Still, there is a certain aesthetic to almost 2-dimensional perpetual motion machines, and to cheap mechanical surfaces that perform energy functions.

If it is cheap to build, someone will do it just to make efficient use of space.

Most mechanical surfaces will probably have to be vertical, which allows vertical movement through horizontal spaces to be used as navigation.

---Quora Message to Jer Ram, 2018/09/30

…

JOURNAL MAGNET PERPETUAL MOTION:

There are no net gains with magnets, although I suppose if you knew how to use magnets to generate heat that might be a source.

---Quora message Mitchell Edwards

…

JOURNAL SWIVEL DEVICE

I have thought of a new design similar to the original pivoting [swivel] device, yet making better use of transitions. I suppose I will post the drawing under the original... [swivel] device.

---Youtube Message to Jer Ram, 2018/10/07

...

[T]he cardboard only moves up in the last fraction of a second of the first movement, in order to make sure the marble makes the sideways transition. Initially it is the [counterweight and] lever creating the motion. With proper scale the sideways motion could take place using a deflecting panel. Likewise, the downward motion of the *cardboard* is only necessary because the lever is flimsy. The downward motion of the marble (lifting the counterweight) is automatic.

---Message to Matthew Egan 2018/10/19

…

I tried buying some metal rods but there appeared to be too much friction on the block of wood I was attaching it to to produce the same effect. So, what I built with plastic is much lower friction than I thought. So, I think at this point if I build it larger it may require a very heavy ball and very heavy counterweight to compensate for the weight of the metal rods, and that means that there will be standard structural concerns for larger builds, which makes it very hard to find the right materials and ways to modify them. But I hope eventually I will find some special way to pivot a metal rod. There is a major concern with the need for stabilization that makes it more complex than a standard seesaw. Not only do I need to stabilize near the fulcrum, but I need it to remain extremely low-friction, as well as providing enough support that the lever does not bend downwards. Maybe I will try stabilizing k'nex with a piece of hardened wire, that might work, but it will look electronic even though it isn't. It would be better to use a single piece of metal for the lever, but I think most aluminum is not up to the task and hollow steel or brass is already a bit too heavy even with small steel balls. As soon as I use big steel balls, I won't be able to support the track properly in the video and I am not even sure how to build a track for a heavy weight, maybe multi-layered cardboard. I still haven't figured out how to make the pivot work if I use a 2-in. ball. And I can't just load pennies with duct tape for the counterweight with that massive a ball, it would be a mess automatically. Even a plastic cup full of pennies wouldn't be heavy enough to counteract a 2-in ball, and it would have to attach sideways so the coins would fall out unless I got a lot of them glommed on the duct tape. Maybe I'll find a new material to work with or find something inventive to do with k'nex. Right now I am finding k'nex are too flexible and it would be better if I could work with metal balls, but I know of no way to do so unless I find some way to make a heavier swivel work. I think someone more experienced could do it very easy, but I haven't seen evidence of anyone else working on this particular perpetual motion design. It's kind of creme d la creme d la creme, but takes a lot of experience to even see it in all it's significance. Anyway, thanks for your support.

---Message to Jer Ram, 2018/10/29

...

Note on Swivel Device: Friction may be reduced with a sandwich of rings or wheels rotating above and below the fulcrum and blocked so as to allow the lever to swivel straight at the correct angle, supported by the sandwich with minimal rotational friction. Ratios may be accentuated by a low-friction wedge-like property

---2018/10/29

[Describing how to overcome the structure and friction problem:] If the effect of the counterweight can be maximized while keeping the lever stable, that is good!

---Message to Unknown, 2018/11/06

...

Like I said, the only inputted energy was for the sideways motion, which is not essential for the back-and-forth motion of the lever. I'm sorry you did not realize that. As many toymakers know, small differences in construction can spell the difference between a working toy and one that does not work at all. So, I am stipulating that some of the very minor physical features present a challenge that can easily be overcome with more careful construction.

In my model, the basket sometimes rotates somewhat, and the 'speed bump' in the duct tape has to be carefully repositioned every time I do the experiment. Also, I am creating the angle in the cardboard track by hand, inconsistently. If the angle in the track is consistent and the basket does not rotate, and the speed bump in the duct tape is more precise and less crude, and the lever is not flimsy but instead shoots back and forth at a shallow but straight angle, and if the basket is located at the correct altitude and the track is sufficiently smooth, and a way is found to support the marble with the back wall of the track while it is moving upwards with an ability to be directed by the angle of track into the basket, then it would work just about the same way, but it would be a fully-functioning perpetual motion machine.

I was simply saying, the mechanics is workable, and here is my evidence. I was not saying that the model was saleable or would make a good scientific presentation, or not exactly.

---Message to Mike Hoath, 2018/12/02

...

So perform the experiment without handling it. ---Mike Hoath

I'd like to, but I haven't found an extremely sturdy lever. The downward motion of the cardboard is necessary if the lever is flimsy and the marble sinks when it enters the basket. The angle of the cardboard cannot be fixed and precise unless the lever does not bend. —Nathan Coppedge

...

The nearest to perpetual motion we will get is e.g. the Voyager spacecraft moving through interstellar space. ---Mike Hoath

No, you're incorrect. This time there is a principle of motion, as the video well demonstrates. - ---Nathan Coppedge

As far as criticisms of the Swivel Lever Device,

There are cases like that [in which significant motion is artificially added], but it is not all of the cases. The Swivel Lever has a worksble ratio around the mid to upper end of a counterweight mass of 2 - 3 X the ball's mass, with a leverage ratio assumed to be an average leverage distance of 2:1. Also, the back wall of the cardboard is used to create a slight wedge action and the lever''s height undergoes some changes from being slightly at the base to slightly at the height of the marble. However, some parallelism is created because the track is also gaining altitude on the leftward motion, so between these the marble might have an advantage as well as the other advantages.

Notice the marble begins on the track and the upper transition occurs automatically. To prove the lower transition, simply realize: 1. The lever is going to be at the same altitude as where it was at the beginning when it returns, 2. Some videos show only a very slight difference in altitude at the trsnsition point. Part of which is simply created by the weight of the marble on the lever, which could be avoided, and 3. The ball is ejected from the basket at minimal effort when no track is present, also 4. If there is more trouble with the lower transition a very sensitive basket can be introduced and the travel distance can be decreased also if necessary to create a shallower transition possibly with slope from the basket itself.

---Nathan Coppedge, YouTube message 2019/02/08

"Secret: The marble need not be stuck to the lever, and may gain as much altitude against the lever (minus losses) = to the original gain in height of the track on the return. If the loss of the marble against the lever plus the gain from the track is less than the gain from the track, the marble will actually return at a higher altitude. So, as a rule, if the slope of the track is more than 1/2 the slope of the lever, a higher return will be possible. This might be true for example if both angles are very very slight." ---Nathan Coppedge, 2019/02/12

"It is thought if the track is a twisted elongated half-donut this may improve performance. However, this is partly outside of my design capacity." ---Nathan Coppedge, 2021–07–16

...

JOURNAL ESCHER NIBW3:

2018/10/28: A modification might use an Escher-Delta type, 2-directional twisting track to accentuate the motion.

—An Argument Defending the NIBW 3

...

JOURNAL TESLA COIL:

If perpetual motion generates energy, the answer is that a Tesla coil would not disrupt the energy unless it extracts it mechanically, electrically, or conductively.

...

JOURNAL ESCHER MACHINE:

The angle of the board acts on a wedge directed rightwards and possibly upwards-directed, microscopically, like 1/10 In gain per 10 inches or less = 1/100 In/ In. vertical change or less, upwards directed.

—Message to Kelen Lukie, 2018/11/07

...

The estimate ftom the article of about 1.1 degrees is in the ballpark. I was just writing about applied perpetual motion, maybe more should join in the fun. Great minds think alike... Google was working on cold fusion recently...

—What is your theory behind Graphene's "magic angle"?

...

Amazing Realization: Recent equations suggest it works based on the principle of *the 1: 0.5 lever and wedge*. This may suggest it works much better when the board working as a track is very thin, meaning the vast majority of mass is creating leverage in the direction of the wedge. Assuming a mass of one, *If the effective leverage exceeds one over the effective mass... perpetual motion with one moving part.*

—2019/05/31, at Escher Machine (Quora)

...

"I may have seen better results with even smaller ledges [than 0.5 in]." ---Nathan Coppedge, message to Judah Yodh

...

The one penny adjustment is only to account for the changing gap in the metal bar, without the adjustment normally therre could be a fault seen with the upper angle of the level due to the fact that the narrowing gap would create a difference in angle between the height of the brackets compared to the actual track of the ball, as the ball is traveling at about a 45 degree angle of the board along the bar which is not a cubic angle. Due to the fact that the brackets are the same, just at different widths, the difference in height is seen to be slight, as the bar obviously is close to level. ---Nathan Coppedge, Comment on Youtube, 2021–04–07

...

"There is such a thing as replication. Perhaps I'm not the optimal person to do those 'first year undergraduate' experiments? I think what most people trained in physics means by a rigorous experiment is one that disproved unexpected information. But in this obscure case that appears to be *barely* impossible to disprove unexpected information. Now you're going to cite me as saying I think the Escher does not work, which is not what I said. There are too many eggshells to dance through as I have said before to phrase things like normal physics. I would much rather post my best simple evidence and leave greater sophistication to someone who knows how to talk like a scientist. If you or someone listening is such a person, what you need for the experiment is a large square board preferably very flat, a way of attaching an angled rail so that it runs at I would guess about 320 degrees or more on the board, extremely solidly attach the rail, and there may be a few other factors like making the rail sloped downward to the left in this case of 320 degrees or more. Then prop up the ends of the board so that it runs upwards, which can take a lot of practice. Basically, it uses a technique similar to a twisted ribbon, with the origin side slightly steeper than the destination side, and the overall angle very similar to 45 degrees but with a crossover between the originally before adjustment downwards slope and the angle of the rail on the board." —Nathan Coppedge, Quora comment 2022–02–09

...

"NOTE: The board is set at about 45 degrees, so the rail is actually close to being right underneath the marbles. Thus, it is unlikely that the marbles are losing altitude against the bar. It is a setup where in many cases the balls will fall inwards towards the camera when they reach the last portion of the rail. This suggests that there is not a significant indentation in the board." —Nathan Coppedge, Quora comment 2022–02–10

JOURNAL SPONGE DEVICE:

[CRITIQUE] In order for this to work, the sponge must be at a lower level than the surface of the water, as it absorbs water. Even so, it would only be a perpetual motion machine in the sense that it would continue running "perpetually" as long as the Earth and sun exists, but not in the sense that it would create energy.

—Legaata, Youtube

I have found there is some water absorption by the cord if the cord is originally completely dry. If the cord acts as a sponge, when one area gets wet the next area gets little wet too, until the whole cord is moist. Then the moisture that reaches the end has no better place to go than being absorbed in the actual sponge. It is not ideal, but when the cord is lifted out of the bowl, it could become completely dry if there is enough heat surrounding the system. It would take a long time, but perhaps it would work sometimes. I definitely observed the sponge getting wet through absorption if not capillary action.

—YouTube Message to Legaata, 2018/12/09

PERPETUAL MOTION BATTERY:

Perpetual motion was invented 60 days ago. If what you are looking for is infinite energy, you may not need a battery, but a perpetual motion battery is not a bad idea, and depending on what you are doing you may be able to call it a battery if you like.

Here is my sketch of all possible energy levels for perpetual motion machines, which can be interpreted physically or philosophically:

Guide to Volitional Energy Ratings

—2018/12/10 (Is it possible to make an infinite battery?)

JOURNAL CONTRIBUTIONS TO V-GATE:

I finally see what you are doing here. 12/2018 (It is unrelated to most of my designs, as I do not use magnets). I thought it was worth pointing that you could have vertical repulsion on the lever end, possible efficiency there.

—YouTube Comment

...

JOURNAL WATER LEVER:

Limitations of the Water Lever: (1) The water lever should be <50% vertically disposed as per usual rules, that is, most motion should be horizontal, (2) I have noticed several flaws, first there may be a need for funnels, which is a major drawback as funnels may often create loss of altitude creating vertical motion, (3) Initially the temptation is to transfer water over the fulcrum, however this is not equivalent to what is normal in my usual leverage machines, (4) Implementing separate water packets has limitations also, as fluid has certain natural limitations.

However, If motion is mostly horizontal, perhaps fluid properties can be used horizontally. The usual limitation here is that (1) without much vertical motion, the fluid will not have much mass, and (2) vertical motion must still take place relying unfortunately on relatively small differences of mass due to the relatively shallow quantity of liquid. Provided that the structure is super-lightweight and involves mobile terraces, advantages might be had through horizontal motion such as by using swivels for the lever holding the terrace or siphon / tubing. Additional efficiency might be had by weighting the mobile end with a counterweight in addition to water, or by using a counterweight on one end and fluid on another.

—2018-12-15

...

JOURNAL "ESCHER LEVER":

SUCCESSFUL PERPETUAL MOTION EXPERIMENT 4

BELOW: LEVER, UPWARDS SLOPED

HEAVY LOW-LEVERAGE COUNTER-WEIGHT OFF LEFT

NOTE: CARDBOARD NEXT TO PLATFORM HELD BY LEVER IS IN A SPECIAL DOWNWARD-SLOPED ANGLE

RIGHTWARD MOTION AT NO COST...

ABOVE: A detail from a photograph of the first Escher Lever experiment (before it had that name).

Design secrets of the Escher Lever to actually help those seeking perpetual motion: 1. The lever is miniscully upwards-sloped in the direction of its tip-- the exact exact degrees count, 2. The lever is bent imperceptibly downwards, creating a relatively flat trajectory for its initial angle, but it is still upwards sloped as shown by the natural return. 3. The counterweight is fairly light, basically some duct tape and a few dimes taped to extremely lightweight plastic toys. 4. Reducing friction is an important factor both in the hinge and where the marble rolls. 5. A possible backward (contingent or perpendicular, away from the camera) tilt was introduced in the fulcrum (support point for the 'lever hinge') about the width of sheet of cardboard. 6. The exact exact height of the 'lever hinge' relative to the twisted cardboard matters quite a lot, I think this is why the experiment was not done before except by me (both my grandfathers were high-ranked engineers is a reason I might have beaten some of the competition).

--Nathan Coppedge, comment on YouTube, 2019/03/21

I have noticed usually there is more of a triangle when upward motion is created. Here, since the strips are parallel, if one strip were higher on one end one way or another the strips would have to be twisted and there would be no natural return. So, an upward directed transition does not explain the motion. Ordinarily there would be a triangle with the base on the left if upward motion was to blame. ---Nathan Coppedge, YouTube Comment to C Mojo 2019/03/22

...

It is partly an illusion of the angle of the camera. I have taken photos from the side that show altitude is not lost on average between the mid-near and mid-far points of the platform, suggesting that all the motion comes from mechanics. I drew lines very carefully comparing similar points of the two photographs---perfectly straight lines, including comparison lines for at least one fixed position shared in the photos to make sure they were at exactly the same altitude, and then comparing the position of the marble. My conclusion was the marble was at or above the initial position when it had moved to the other side of the divider. To explain why this is possible, one must keep in mind the near motion occurs because of downwards slope to the left. This motion compensates for the upward direction of the first extent of the platform (far side) leading to the right. The far motion has been shown to occur mechanically even using an upward-directed lever, due in a very special case to the special twisting in the cardboard on the side of the platform and the support from the side of the platform. The angle is close to horizontal, but it must be upwards-directed slightly also, as shown by the natural return motion. Since the far motion occurs naturally and is upwards-directed, and the near-motion occurs automatically and is downwards-directed, it could be reasoned that it is a secure principle for perpetual motion, assuming it is not necessary to lose more altitude than is gained, and assuming the two sides of the platform are at equal altitudes OR that twisting in the platform may be used to cause inward and outward motion as I have observed, which can be equally good even without equal altitude at every point. Although this sort of operation is inherently baffling, experiments have shown that it is indeed workable, at least and perhaps beyond the point of getting four-directional motion on the horizontal from rest.

—Youtube comment by Nathan Coppedge, 2019/10/07 SUPER FUNCTIONAL PERPETUAL MOTION IDEA

...

In principle, if the marble can move on four sides of a quadrilateral from rest, this suggests something similar to moving on a level surface at no cost of energy. Although it might be argued that the marble did not return all the way, it cannot be argued that it moved in a spiral, and effort was made that no energy was added to create the motion. This suggests motion that nets zero change in altitude but that occurs from rest. That is very similar to a definition for a perpetual motion machine.

Motion along four sides of a quadrilateral does not normally occur from rest, that is all I was saying.

—Nathan Coppedge, comment/124021408

...

"On Jan 15, 2020 experimentation proved nearly impossible with thin three-thickness cardboard. For a time it seemed to me like the design didn't work. However, this seemed to result from paranoia that God was out to get me. Subsequent experimentation with the Verticsl Lever (a different design) showed similar principles to the Escher Lever were still workable, it was just somewhat lucky to have noticed the correct configuration so quickly. The conclusion was, the devices have at least something similar to operability in certain obscenely precise ratios, and finally it seemed conclusive that the law never changed." ---Nathan Coppedge, January 16, 2020.

...

SCARPA'S PENDULUM:

Alternate, New experimentation shows least resistance with very small, heavy tethered orbital ball taking close to center position supporting a large, lightweight loose outer ball and as small a bowl as allowable, perhaps even some type of grooved spiral base. May show new use of ball bearings, with efficient load-carrying, for example if shifted balls can create alternating directions of slope over long distances when wedged tracks are placed over pairs of whole units of this type.

---2019/05/10

NATURAL TORQUE DEVICE:

Concerning a concept similar to my earlier Natural Torque Device (practical insights on this device occur rarely, but the ascending vortex idea occurred earlier with the first version of the Perpetual Motion Flying Machines graphic from 2019):

"Interesting idea, I'm sorry I did not reply immediately, I did not even have the recent edition of messenger installed. I suspect you have reached the so far rare stage in perpetual motion design where it is time to ignore the videos and focus on the extremely clever. There are merits to your design, it has many clever aspects, but I suspect as long as it is just magmets it may not be 100% workable. The reason is magnets do not really add energy, they just provide cushioning like a pillow. However, a spiral-type arrangement may work with the Natural Torque devices, but it has not been attempted yet. One of the problems with vertical spirals is they must involve vertical motion, which is not possible unless the disks rise or fall. However, this implies a circle or purely ascending motion which is normally impossible. Magnets, I have noticed, do not tend to help with ascent."

—Facebk Message to Mi-tree Ray, 2019/05/28

...

"Natural Torque has been restored as my best perpetual motion, but I have found it is more efficient with a shorter-distance counterweight and no side lever." ---Nathan Coppedge's Top Perpetual Motion Machines, 2019-08-22

...

" I just realized Dec 1, 2019 that if you put a carefully weighted wheel on the short end and an upper track at that distance above where the long end is rotating in the leaping abyss video, the result is 360-degree automatic motion." --- Statement on the Perpetual Motion Links page, December 1, 2019

...

REPEAT LEVER 4.2:

Equation [from earlier]: Assuming ball = 1 with variable application, and long end has additional 1 constant application, and counterweight located on shorter end, and counterweight is designed to direct ball on opposite end up slight supporting incline before ball applies leverage, Unified Counterweight Mass Formula = Min Lvg + 1 > (Max Lvg / 2) + 1.

The undesired alternative to the inward spiral wall design requires modifying the equation to correspond with 1/2 counterweight mass yielding in that case an unworkable equation, or using a design with less horizontal angularity in the lever (for example, having the ball run in a narrower ring underneath, which may be used with an outward spiral for the counterweight). —Nathan Coppedge's Top 10 Perpetual Motion Machines, 2019/06/13

JOHN WILKINS:

That is a concept that has been investigated, I believe it is called the John Wilkin's perpetual motion. Perhaps with a combination of a regular magnet and an electric magnet it would work. I suppose either no one thought of this, there was insufficient return energy, or something like that. Remember in this device the magnet has to be very strong creating a need to decrease magnetism on the return. Maybe if the track is used as a lever acting on the height of the magnet sometjhng interesting would happen, if an advantage can be had on changing the height of the magnet when it is scted on. This msy involve a loop in which the magnet returns to a near position, then toggles the magnet out of range, plus part of the magnet is an electromagnet with batteries switched in and out. I am not even sure if electrically powered magnets exist, but there are magnets called electromagnets. However, if they are electromagnets a sustainable energy would have to be found for them stemming only from the motion of the ball, meaning there would have to be permanent batteries. Another possibility is a machine that uses very slight slope then uses steeper slope to pull the ball away from the magnet. This might even be used in succession. Possibly some of the problems are similar however. Or maybe a setup where a heavier magnet pulls a ball through a pinch-wedge with a smaller pushing magnet on the pinch wedge would work. Combinations of powers of magnets in multiple directions is a relatively good bet with magnet devices.

---Message to James van Pelt

...

RELATIVE MOTION MACHINE:

I imagine I interpreted this from a retarded boy who thought it was the greatest idea. However, I have never seen in in writing or in diagram form.

It is also possible the idea was repeated by a retarded boy who received instructions from someone gifted.

The way I imagine it, two wheels spin countrary to eachother. The difference is used for exsmple in a spiral fashion to resch high speeds.

However, this concept by itself does not provide a volitional-mechanicsl means for motion.

BUOY DESIGNS (VARIATION ON FRANK TATAY):

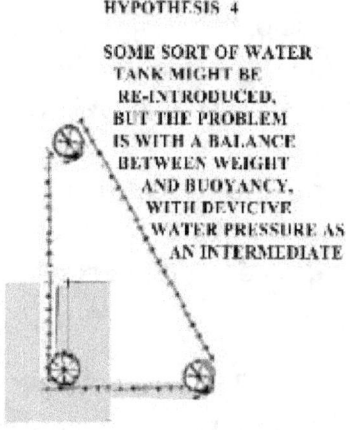

HYPOTHESIS 4

SOME SORT OF WATER TANK MIGHT BE RE-INTRODUCED, BUT THE PROBLEM IS WITH A BALANCE BETWEEN WEIGHT AND BUOYANCY, WITH DEVICIVE WATER PRESSURE AS AN INTERMEDIATE

ABOVE: A variation on a buoy design (the variation is from me in 2005) aims to maximize vertical buoyancy while reducing resistance.

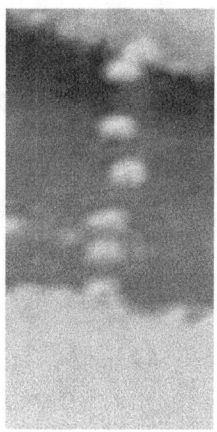

However, it has been observed that buoys lose energy when placed in a vertical column. Although I had hope the falling buoys would help, this design is very hard to build. The design might benefit by additional chain attachments for the buoys, which might allow spiral behavior. The buoys appear to move best independently while still attached to the chain, but this may require a massive scale to compensate for the backward chains of the buoys entering the tank from below.

An additional possibility is suspending the buoys within a substance thinner than water so as to reduce the entry resistance. However, this requires ultra-lightweight buoys of sufficiently small scale and the right medium.

—Pre-Perpetuum

...

'ELON MUSK' PERPETUAL MOTION MACHINE:

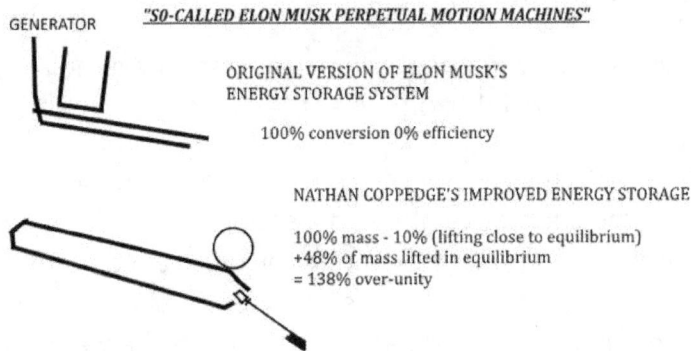

Basically, if the weight attached to the lever is more than 60% of the weight of the ball, additional weight added counts as over-unity, assuming low friction and 50% return of mass-energy in the ball while it returns.

DETAILS:

Basically, (FIRST METHOD) a heavy weight on train tracks at an angle has 100% conversion minus friction, which is pretty good if the friction is low, or if the input energy is cheap... But this is not the same as 100% efficiency, it is more like 0% efficiency. which is pretty good if the friction is low, or if the input energy is cheap... Although, (SECOND METHOD) using a *weighted lever* to lift a weight to the top of a long track may be more effective.

(Other notes):

Also, this method is overall very cheap considering how easy it is.

(Even the first method) is much better for example than lifting weights on pulleys.

I estimate the second method if it uses a long lever with the counterweight, and a short distance for the lighter end, could create net-over-unity effects when the whole energy system is considered, as only close to equilibrium would be necessary to lift the weight, and part of it could be considered as leverage rather than mass.

Therefore, 100% mass - 10% (lifting in equilibrium) + 48% is about 98% efficiency = 138% over-unity minus friction.

That is at least 38% more efficient than not using the counterweight, and incidentally creates free energy if used effectively.

The advantage here would also be the great speed with which stored energy could be created.

What is the most cost efflctive means to store energy using today's technology that has potential for mass adoption?

"Only problem is it's too slow." —Elon Musk (?)

Then scale up! —Unknown

Modification 1:

I have considered maybe the support on the return reduces the 100% to 50%. So here is a solution to that.

Use very slight change in altitude so that the ball can depress the lever when it returns to the plate using most of it's mass.

Now take the difference between the total mass and the amount the lever can lift, which is something between 50% and 99.999% of the ball's mass if we assume the lever is heavily weighted and that the ball can depress the plate automatically.

Now, look at the over-unity rating: it is equal to:

50% -10% + (50 to 99.999%).

Now, if the lever is weighted to the equivalent of above 61% of the ball's mass, we STILL get over-unity in ideal cases. And then, we have up to another 38 units we could add to reduce friction.

If all energy can be extracted from the falling weight, these new equations give us about less than another 50% on top of that, which would give us between 101% and 188% over-unity. In practice however, the value is more likely to be 101% to 138% identical to the value given earlier with the less conservative equation.

...

—Nathan Coppedge (June 30, 2019), 'Elon Musk' Perpetual Motion Machine

...

Two sets of carts moves back and forth like a gravy train (on two close parallel tracks). However, equidistant from the middle, two of the trains have installed what I call an 'inverse difference pendulum' which is a rod, slightly weighted on the end, which may be caused to move in an arc, with the extremes of its motion blocked at the base, in this case with the range extending to about 45 degrees or less to either end of the vertical (some modifications may be applied to this, for example, a narrower arc with a bar attached above the lever to make automatic movement of the lever easier). Now imagine that next to the bar on one side of each track is an elevated track designed to support small wheels. The wheels are attached to one side of the bar on that one cart for each track corresponding to the one side for each track that has a special platform for the wheels. Now, imagine that both pairs of railroad tracks have a high end and low end on opposite ends compared to the other railroad. Now, it can be designed so that when an inverse difference pendulum is unsupported and the opposite inverse difference pendulum has already been toggled to the supported side, the unsupported difference pendulum can have sufficient difference to raise the opposite train as its own train is being lowered, since the resistance of the opposite inverse difference pendulum is reduced due to support from its own special raised side-track. Now, when the unsupported inverse difference pendulum reaches the end of the track where its own train is blocked, the momentum from both trains can be used to toggle both inverse difference pendulums since the shift always occurs in the same circular direction for each train on either side of the track, thus toggling can occur in the same way consistently for both trains on any given side, except that one side results in toggling into support, and the other side involves toggling out of support, but since the railroad track doesn't itself tilt, toggling is always the same on a given side for each railroad. Now it is clear if we compare the cycles, there is always an advantage of displacement except at the very short transitions which use momentum, so that the train beginning higher is always given an advantage in lifting the opposite train that has the supported car. Since this arrangement is designed to repeat every half-cycle there is no real need to complete a full circle before the device can reset itself. Now, that is a fairly simple design that might work when you think about it! It probably just requires weights that are sufficiently heavy with sufficiently small change in angle to allow the motion to take place. An added detail is that it might be best designed in such a way where the supported inverse difference pendulums lean against a kind of vertical wall, allowing the mass to be supported closer to the angle of the bar.

—Nathan Coppedge's answer to What are some new cutting edge energy projects?

...

JOURNAL BRACKET LEVER:

This may be a record for time from design to construction. It felt like several minutes flat for the whole process. This design and variations are now highly recommended.

Alternate Modification July 25, 2019: Counterweighted lever pivots on a spiral support. Opposite end has wheelie. Wheelie is designed to gain ground against slope using counterweight if it is shallow. When it is slightly steep it moves downwards.Now, if it can be arranged that the track is very gently counterbalanced, and the upward end of the lever's motion causes the track to tilt downwards for example by pulley against a lever, then the cycle may be perpetual.

…

JOURNAL DESIGNS BASED ON CHRISTOPHER ANTONIOU:

"Perhaps the first workable concept for horizontal wheel perpetual motion, a design by Nathan Coppedge based on a different design by Christopher Antoniou: My solution if you will hear it is to keep the [horizontal circular track] use 1/2 the inner track with slight upward grade, remove the other half of the track entirely except an outer rib for structural and aesthetic purposes, then mount the heavy ball on a higher pivot using a tether of sorts on the short end with a lighter counterweight rotating on the longer end of the pivot. My sense is this would work on momentum if a guide track were used allowing momentum to push the ball up very slightly at the entrance to the upward grade. The ball would gain momentum in both halves of the cycle, due to 1. Sufficient leverage applied during 1/2 mass * distance, and 2. Heavier mass relative to opposing leverage when the support is not present. The near-perfect wheel shape would facilitate the motion in this case, which is an effect I had hardly seen before. In fact it is tempting to try spirals however I think this result is much better. This is an exciting opportunity for those that want to build perpetual motion. Added detail: an upward-directed v-shape in the track entrance may facilitate." —Nathan Coppedge, October 3, 2019

...

JOURNAL BALLOON-BASED:

Best results a little shallower than 45 degrees from top (easier to climb), about exactly 5 degrees vertical slope to the left, angle of board is directed slightly diagonally from mid-left of board, in line with start of balloon weight. Extremely low friction, balloon is a very large fat coin shaped balloon containing I guess 100% helium from a dollar store type place. I bought ten balloons but so far only needed one. Thumb tack was way too light a weight for these balloons.

--Nathan Coppedge, Youtube comment, November 30, 2019

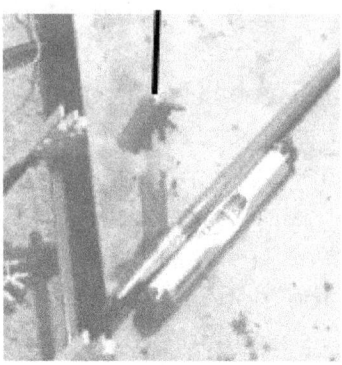

For one: Below, the piece of plastic appears to be moving ahead of the balloon. If the plastic is ahead of the balloon, this suggests very limited wind, which proves it works as a perpetual motion machine. If the plastic is not ahead, why does it appear ahead? --Nathan Coppedge, December 3, 2019

Recently I have been distracted with helium balloon experiments which were a long-term goal, and I was somewhat proud of myself to finally do experiments using helium. I'm not completely sure helium works without wind, it seems possible but I just don't know. I think it could make logical sense that vertical force from helium could cause slow motion along the path of least resistance moving upward and mostly horizontally, however it seems likely in both major balloon experiments some form of wind was a significant factor. If it was the only factor, then the balloons do not work with their current principle. But if wind was not the only factor, then it seems likely these designs for balloon pmm's both work, if a bit slowly. There is also a factor that helium balloon perpetual motion seems to be affected very easily by extraordinarily small differences of mass in the dangling weight, such as fractions of sections of scotch tape. My intuition here is that maybe this indicates that it can work without wind, but is unlikely to be discovered by most of the 'insensitive' people doing experiments. I still think it seems unlikely that sound was the only factor on the reverse 1/2 mass X distance balloon experiment, which in effect suggests that both major types of balloon pmm's do work. But there have been some cases where the laws of physics seem to turn against me as if someone tweaked the levels of resistance, so it is hard to trust my instincts when the world seems set against me. At least I can say that it is possible to have a large-scale indoor sport called Perpetual Motion Landsailing, which would offer very similar benefits to perpetual motion, only using small amounts of wind, but still acquiring the advantage that the weight of the vehicle could lower the balloon in spite of the vehicle's acquiring altitude with very small input using the principles of 1/2 mass X distance.

—Nathan Coppedge, Youtube comment, 2019–12–25

My current guess is the balloon PMMs work but require obscenely precise weight ratios and obscenely low friction. ---Nathan Coppedge, Comment on youtube 2021–06–12

...

JOURNAL ON WORK OF BENJAMIN BOUGIE:

Advice for Benjamin Bougie's trapped low-friction magnet column:

You have an above average idea. However, a vertical wheel of this type won't work: it will need energy added.

My sense of why this is is that when the longer end is lower at the middle point it loses the ability to lift itself. Using more than 180 degrees poses problems for a straight element. Also, a slanted wedge cannot be used to reposition the *outer* end, because that end begins in an upward position after the intended motion.

I know this is counterintuitive, it is the problem of assuming different centers.

It is an important stage, you have encountered several important stages here. You might want to look at the work of Christopher Antoniou, you both are at a similar level.

My intuition here is you will need to go more horizontal. It's not completely worthless.

I should add your video made me feel better, I have been looking for other authentic perpetual motion hobbyists without much luck. This may be a turning point.

—Perpetual motion magnetic force lever

...

JOURNAL ON "DR. PERPETUUM":

This one inspired me a bit. What if you had 0.6 weight pendulum supporting a 1X ball rotating horizontally, partly supported on a platform? What if the ball was hinged to rotate about the center, and the bowl was spiral? --Nathan Coppedge, on Perpetual Motion Research #169

...

JOURNAL VE PROJECT SPOKES LEVER ("UNBALANCED" LEVER):

A variation on this might use a fixed balance as a propeller weight. Another idea is to use mass versus distance with one end of the lever being shorter and heavier, and the spoke element working the same way but with additionsl weight slightly forward on the long end using momentum. This may require positioning the heavy end at avery slight bracket angle backwards. No one prejudge, this design has similar principles to designs that have shown natural momentum without cheating. Worst comes to worst, sdd an element of the design that uses 1/2 mass X distance with low friction. Another idea is to use horseshoe tubes with weights that float above where the opposite water is. The net height permits motion because the opposite buoys are sent upwards along the curve of the tube. A variation on this is hollow helmet- or pole-shaped buoys that allow extreme gain on buoyancy. A promising idea might be to use thin drooping chains with buoys, perhaps sealed in a plastic tube, if poles don't work.

—Nathan Coppedge, comment at: VE Project: Unbalanced Spokes Lever. Perpetual Motion Machine

...

JOURNAL AREA OF DIFFERENCE DEVICES:

Based originally on a video at: Oval & Balance

Applying ball to area of application might result in a turning wheel.

Might work horizontally using natural transitions, perhaps an inner and outer spiral at varying altitude from varying area platform bumps.

JOURNAL DOLF PERENTI ORGANIC ESCAPEMENT #2:

(Assumed typical escapement) How long does it stay moving by itself?

Wondering the same… It seems to me with a heavy weight at short distance counteracting a lever it might be perpetual, however this might require reducing resistance in one direction such as by using an additional rolling ball partly supported by a fixed element, and letting the ball operate the lever.

—Organic Escapement #2

JOURNAL WATER BELLOWS:

If it breathes through the top the entering air would still require displacement, and would be displaced by water.

JOURNAL SMOT:

Finally commenting on the SMOT, usually not much to say here, magnets can be problematic.

"Recommendation: use larger ball with mass > 1X < 2X the magnetism expressed in mass. This can allow the ball to pull away even without help from magnets. Another recommendation: I have found in my modular magnet concept using similar principles a highly vertically angled drop with some deflection can assist when combined with momentum. Perhaps this would have good effects with a larger ball in your design."

--Nathan Coppedge, -Comment at Brass wood and glass MarkIII SMOT

Note: As motionmagnetics (youtube channel) may have noted, the key is to pull away with some altitude gained, so a combination of weak sideways-directed magnets acting at a distance on one side with a wedge effect, perhaps directed at some type of V-trough may be effective, or finding a similar way to move very heavy balls using slightly stronger magnrts, perhaps with very shallow change of altitude. They may have to be permanent pull magnets if such were possible arranged to be closer and closer to the V-trough near the top, but with < 1X the mass in pull strength and with the drop being less than the height of the chain of magnets and with horizontal deflection. If you can prove altitude gain from rest you're golden. Like I said, this may require a short chain of weak magnets with very little change in altitude, and it may require permanent pull magnets.

CSABA HORVATH, PERFECTION OF THE SMOT:

See
link: https://www.youtube.com/watch?v=LXs8SYWwkl

M (https://www.youtube.com/watch?v=LXs8SYWwklM)

Csaba Horváth has the best variation on the SMOT. I think it's inexplicable that others have not reproduced his work. If you want to replicate his experiment, follow his advice: "There are 10x10x40 mm Neodymium magnets on both sides of the steel rail. The magnets are oriented one side to the south and the other to the north. The magnets are placed closer and closer to each other on the steel rail, making the magnetic field that attracts the steel ball stronger and stronger. At the end of the steel rail, even in the strong magnetic field, the ball falls down due to its own weight. Here, in the end, a precise setting is required, but with some smaller Neodymium magnets this can be solved. You can build one too, and please submit the video." However, Horvath claims it's just a toy

JOURNAL MOTIONMAGNETICS:

"Adding to earlier comment, a goal would be to pull only leading up to the drop point, drop it at least 45 degrees, not symmetric lead up and follow through, use mass drop with deflection (wedge = 1.25X multiplier) and lower track further from lead-up magnets with projection from momentum away from 45 degree drop, yet suddenly flat. Then if possible draw in with the next magnrt just enough to allow next drop at same or greater energy level; the key may be weak magnets positioned close to track in short yet separate intervals, letting 45 degree angle do much of the work. For this I have found less of a bar but weaker over shorter distances including shorter width more effective. Additional secrets keep very base of drop points a bit shallow (still downwards-sloped) to extend earlier mass-wedge effect. Position magnets so they are not located immediately on the lip of the drop, but eomewhst earlier to reduce magnetism. Allow the pull from the magnet to sling ball barely over ledge with thst momentum translating directly into the mass-wedge effect. I saw some results already slthough interaction between ball and magnets numbed this effect in series. It was close to working in series with consistent altitude even in my crude attempt, however the magnetic effect on the metal seemed to wear down the results over time. Another idea, sling over a hump with some but not much resistance before wedge, allowing more efficient sling with gradual but still less back-pressure when the magnet free falls, use magnets and angle of arc exclusively to permit gradual altitude gain which could be used to ultimately roll marble to beginning. It may help to use long-range weak magnets that wear off after three inches or less but with semi-significant drop. Earlier reference to shallow just meant the slope leading up to the next magnet begins downward-sloped to help momentum."

---Nathan Coppedge, comment at Magnetic Gravity Gates

Journal Aldo Costa Wheel:

Vertical wheels tend to suffer from the vertical wheel problem related to lifting with equal mass to whar is lifted.

The secret of the big wheel if there is one may be that the lower weights are pulling the upper middle weights slightly horizontally.

Another possible method is imbalance occurring solely in the bottom half.

Raymond's Lab No Back EMF Concept:

Maybe try a two-node bar above and below fulcrum spot, may cause vertical oscillations. I suspect this device requires energy input.

Buoyancy Lever Without A Counterweight:

Combining multiple machines is on the far fringe even for my thinking partly because I have found no way to get additional efficiency with multiple machines combined. The intuition is it would imply adding a further efficiency factor which I do not know about yet. Perhaps two or three levers or buoyancy without a counterweight (Nathan Coppedge Oct 17, 2020)

Yuri Yurin Youtube channel (Юрий Юрин)

https://www.youtube.com/watch?v=vkcFRLQiVGk

"If it works I am guessing it uses an asymmetric method with some verticals. With the first device, it seems possible it is using volitional mechanics principles, however, if so, it likely uses a half-spiral for the repulsion, and some type of wedge-knocking to attain altitude for part of the return, in order to gain mass advantage on the magnets at the beginning. But in my experience with this type, the altitude-hit is not sufficient to help the magnets. Maybe with spiral supported rising motion. The repulsion would have to be > 0.5X < 1X the asymmetric mass or magnet on the wheel, with the wheel rising up and down and the asymmetric mass supported by a rising spiral wedge. There may also be other factors. On the original design, perhaps with some additional mechanism like a counterweigjted balance for the whole wheel this would be doable, however, this suggests the device is not 100% different from my designs if it works. Although my designs are not patented. Another possibility is he got lucky with weak magnets like in my magnetic gate experiment, but I have not seen many examples of full cycles with that method. Maybe it is possible, but if so he's a rarity so far in history. If his device works I hope it keeps working when he runs it, but I suspect he is hitting a switch for a motor." --- comment by Nathan Coppedge

Joao Benedito Youtube Channel

"If this works, it is probably.on the principle of gaining repulsion from the spiral wedge and dropping a mass at the tip of the spiral that is slightly heavier than the repulsion. Still, the way it is pictured this would result in rotation from inputted energy alone, which suggests it is in a class with the fraudulent magnet motors. I would like to know otherwise, but the short filming suggests the device will slow down probably. Youtube allows long videos, post a longer video if it works, showing that there is nothing plugged in and nothing under the table, and no batteries with the magnets, more bare-bones. This video makes me wonder if you are telekinetic. Vertical wheels don't usually work, maybe you are an exception. I would like more information on how the spiral helps. Photons? Solar? I believe in perpetual motion, but in this case I am not sure. How is it so different from other models? Is this a mixture of repulsion and attraction? If so, and it works, be sure to preserve your legacy as other inventors may have had a strange dislike for this method, you may be unique, the extra umph of cleverness. Maybe it works after all. Still, be sure you run it for more than a few seconds." ---Comment by Nathan Coppedge https://www.youtube.com/watch?v=K73R6dMJRCI

Repeated Deflection Device

"Interestingly, it mostly seemed to happen only on Sept 1, 2020 so far not because I'm joking but because I've had a hard time setting up this experiment. In theory if physical laws are true, the experiment should still work in about exactly the right configuration. I suppose I should try it again sometime and see if I can get equal results. Keep in mind, it required a lot of trial and error to get the diagonal member in the right position. A high degree of balance with somewhat heavy weights and high tensile strength seems to help." —Nathan Coppedge, November 28, 2020

"Why does it oscillate? Either A. Because of Brownian energy in other words, vibrations and movements in the air and the floor which are exaggerated by the length of the lever, or B. Leverage acting as resistance against a specific amount of counterweight, which is periodically reduced by contact with a diagonal bar appearing at a certain length along the lever, or C. Some combination. or D. Unnatural torque acting through the angle of the weight on the end of the lever temporarily reducing resistance by creating sideways action. I now suspect a mixture of C. and D. may be the answer in this case because I have not reproduced the same result using straight up-and-down configurations of the weight on the end of the lever. [Notice the taped weight on the right is tilted backward in the image, leftward from the top of the fulcrum]." —Nathan Coppedge revised January 7, 2021.

1st Successful Module Experiment, 'Cat-Trap' Device:

CAREFUL SHOES HAVE BEEN ADDED TO THE RIGHT POSTS AT THE CORRECT LENGTH TO SHOW THE BASE IS LEVEL AND THE RIGHTWARD MOVEMENT OF THE BALL INCLUDES A GAIN IN ALTITUDE!

Edmund Scarpa, a math teacher / professor in Connecticut, seemed to say: "Well, you cannot get higher from lower altitude. You can't. What, you don't believe me? Try it!"

"The point I was making with the levels is that at the beginning the slope is upwards, and at the end the slope is downwards. So, if the right side is shorter, theoretically tilting to the same blocking point would actually result in net upward motion with the correct counterweight on the right side, since we will assume the track is straight." —Nathan Coppedge 2020–11–30

"It seems to be true (e.g. gaining height) in one unit, but is harder to reproduce over two units due to the exact starting height of the lever. The low position of the lever may be low enough, but the high position is close to being prohibitive to movement in my construction." —Nathan Coppedge 2020–12–01

"The insight is that 1/2 effective mass (in other words, the principle of horizontal rolling) can apply on one end of a balance when the masses are close to equilibrium." —Nathan Coppedge 2020–12–01

"One of the problems in my device is the rods used for the lever end are extremely bendy, which has the effect of reducing momentum before there is even a downward slope. That makes experimentation awkward." —Nathan Coppedge, message to Angus Skinner

Later Modular Experiments showed something similar to perpetual motion: "The last video is not precisely level, there is some upward motion after beginning on a ramp, not usually net. What I suspect is that if the difference in altitude is shallow enough, since energy is going in from a counterweight or more than one, since the ball is acting with properties similar to a wheel it will continue moving so long as it has a sufficiently strong altitude advantage at it's midpoint, and so long as the counterweights are strong enough to create upward motion within each module. However, this depends on the counterweight being less than the ball's mass X the maximum leverage." — Nathan Coppedge, Quora comment 2021-11-29

Reverse Escher Machine

"What is encouraging is I have found the window is much larger when the panel behind the ball is raised to about 45 degrees, which makes sense because this reflects increasing momentum from the backboard. I will add to that, the idea that the 45 degree angle increases losses of altitude doesn't make sense, because if you look at the change in angle of the reverse Escher bar in this case, it goes from a slightly stronger inwards slope to a slightly narrower inwards slope, yet the level shows stronger upwards slope and a much greater ability to create motion. The principle is that the backboard has about 50% momentum, while the wedge has about 125%, if you multiply them together you get more than 50%, but when the movement is close to horizontal the resistance is close to 50%, so the two effects on momentum together can actually in principle move the ball upwards from rest, but only very slightly."—Nathan Coppedge, message to his Dad (January 9, 2021)

"[An] interesting case might be a wedge with the typical wedge efficiency of 1.25 acting on a ball which is set at an angle to apply 0.66 leverage on itself. Multiplying 1.25 X 0.6666 produces an efficiency of 0.8333 which when applied to a mass of 1 might produce upward lift enough so to cause upward motion of the ball by its own weight. I call this the Escher Machine! In perpetual motion energy reduces to: Heavier mass range / Lvg ratio + 1 * 100% The Escher is an unusual case that requires its own separate equation. Preferred Equation for Escher Machine: 0.49 X 1.25 > 0.51 = 0.6025 - 0.5 + 1= < 110.25% over-unity [non sic]" —Nathan Coppedge, June 10, 2021

"After much experimentation it was looking like comparing equal balls on opposite ends of the guide wire resulted in downwards slope, which I considered bad. However, in a recent experiment (July 28, 2021) it was shown with tweaking a significant portion of the motion measured the same way would show significant upwards slope in the direction of motion. My conclusion was although it was very tempting to believe the experiment failed, in fact, under the correct conditions the experiment was a success.

Since the ball moved from rest, and the second portion of motion was clearly upwards-sloped, while the initial motion might not even be as long, this suggested that even if the slopes were inverse of each other (they were not designed to be), then some natural upward motion certainly took place initially from rest." —Nathan Coppedge (August 7, 2021)

Double Lever

I was not assuming anything different from Archimedes or Newton, simply compound efficiencies involving the same principles. This specific case is not a very good example, usually I use a supporting track for the weight that is being lifted, and lift it to a position where the track ends, so the weight is then able to lift the counterweight since the limited resistance has been removed. That case is less dramatic in some ways however, because in that case the upward slope is usually very small. This is an attempt to find efficiency of a similar kind without the supporting track. —Nathan Coppedge, comment responding to Nick James

Csava Horvath Youtube Channel

See
link: https://www.youtube.com/watch?v=LXs8SYWwklM

Csaba Horváth has the best variation on the SMOT. I think it's inexplicable that others have not reproduced his work. If you want to replicate his experiment, follow his advice: "There are 10x10x40 mm Neodymium magnets on both sides of the steel rail. The magnets are oriented one side to the south and the other to the north. The magnets are placed closer and closer to each other on the steel rail, making the magnetic field that attracts the steel ball stronger and stronger. At the end of the steel rail, even in the strong magnetic field, the ball falls down due to its own weight. Here, in the end, a precise setting is required, but with some smaller Neodymium magnets this can be solved. You can build one too, and please submit the video." However, Horvath claims it's just a toy

XX0RU Youtube Channel

Perhaps the greatest master of perpetual motion wheels so far appears to be this guy on the xx0ru video channel. It is my belief this concept may work due to the internal balance of the switches on either side of the fulcrum.

Later noticing the potential important properties of combining imbalance with equilibrium made this one stand out. Video from September 26, 2013: Note that I did not notice this channel until February 17, 2016, but I see no real reason to dispute the dates. Perhaps the greatest master of perpetual motion wheels so far appears to be this guy xx0ru. It is my belief this concept may work due to the balance of the switches on either side of the fulcrum. I believe this guy called it a 'puzzle to solve'. If you watch the video very carefully, the exact point at which the wheel would have to be unbalanced is actually behind the final test of the midpoint of where the imbalance would have to be if the wheel were mechanically cheating. Also notice that xx0ru does not add any energy to the system from the outside, unless perhaps he uses rubber bands, but I am not sure of this. The device also does not appear to be very electronically imbalanced, because the xx0ru is able to show somewhat natural backwards motion, indicating there is no ratchet mechanism. This mysterious cypher of 'puzzle to solve' may have also meant 'do it yourself I mean it too much'. Maybe it is not a good sign though that this channel also posts on Russian asteroids and UFO phenomena. But I can see it's possible that he reposted others' videos on those subjects. Then I thought of the term 'zoro sanctum' which got me to thinking this guy is an alien… It's just the top ones and the furthest side ones I suspect that add the energy.

…

'Towtops' Youtube Channel

Not very interesting link present at: https://www.youtube.com/watch?v=SFqkg8hsM6U

A better design is simply to chain the right buoy from the left, and allow the right buoy to be extended by the upper end of the motion. The bottom buoy doesn't matter and is not restricted by having moved inwards. It may also help to have more arms, at least eight and up to 32 or more, though this may not be necessary.

…

Lighter Ball Moves Heavier Ball Longer Distance

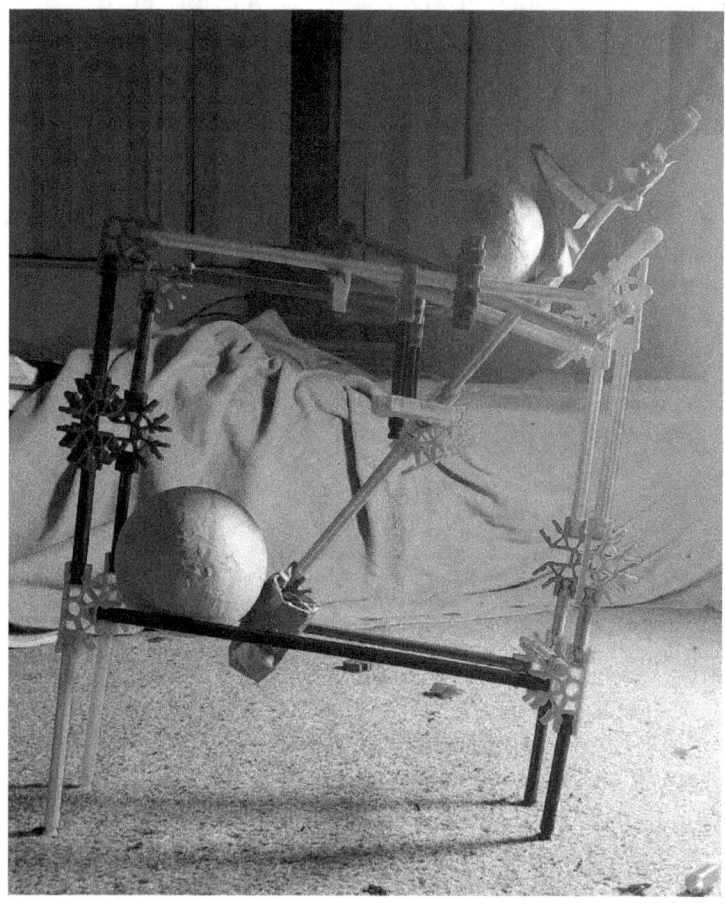

It's extremely difficult but possible. Requires quite a lot of luck to get working, but in theory physically possible (I'm not one to win at gambling, but you know what I mean). I'm extremely experienced, and sometimes had to try repositioning what's called '5 perfect times' or so before it would even look like it moved. But then by perfect time '6 to 9' it can seem close to the above video, if built correctly. Note that there is a specific wedge effect incorporated in both the upper and lower tracks, with the upper being only a slight wedge, and the lower being a pretty dramatic one. I found the wedge is one of the factors that made a big difference. Other factors are the exact slope of the rug and angle of the scaffolding, and the exact angle and position of the 'reverse wedge' cardboard thing attached to the top of the lever. Also important are the numbers and types of attachments to the lever, and the fact that there are one or more pennies attached by tape to the bottom of the lever. Hope these comments help. It does not seem to be impossible, because I have done this experiment many times, but sometimes it falls over if I leave it alone. I found the way the structure leaned was helpful for making it work, although I am not always sure this is the case. It might actually work better with a level floor, but I am not absolutely sure. Also, it is VERY difficult to get it to work as well as it does in the video, but with a lot of re-tries you may be able to make it work better than half as well, if you know what to do. A lot of the adjustment originally when it first started working were with the EXACT distance and POSITION of the central pivot, and the EXACT angle of the upper platform, the exact width of the wedges, and exact angle from above of the central hinge. If the central hinge is not PERFECTLY perpendicular to the outer structure of the device, it never seems to work, but when it is then you can adjust other things earlier, and after adjusting the hinge (fulcrum) IF it is at correct distance and position first, then when you make it perpendicular there is CHANCE to work, if everything is the same as in the above video, OR in some other cases that I don't have experience with. If you are just starting out with learning about your model it can help to test the device with one ball at a time to make sure it goes in the direction it would normally go when there's no resistance from another weight. Sometimes there are cases where the lever is too heavy to allow the ball to follow a slope without resistance from an opposing weight, which is very bad. All of the devices with nifty properties overcome this hurdle long before they become demonstrations. Except in this case it is is a bit tricky, because there is a kind of four-way resistance which is only overcome through properties similar to a balance, use of specific angles, and the addition of the reverse wedge on the top of the device, and special weight ratios. I encourage you to keep trying, or if you have trouble, there are easier experiments like the Successful Over-Unity Experiment 1 which can be more fun because it involves building with cardboard. In some ways the

Over-Unity Experiment 1 is closer to a perpetual motion machine anyway. This particular experiment above was just a flashy thing I did because I had the skill, it's not exactly introductory material if you can imagine a linear process of development for these things. Still, if you can do the above experiment you'd be getting in at a high level at least from my standpoint.

—Nathan Coppedge, Comment to Bhuvanesh D on youtube

LIGHTER-THAN-WATER DEVICE

Here is a water wheel which may show some potential: The Lighter-than-Water Device:

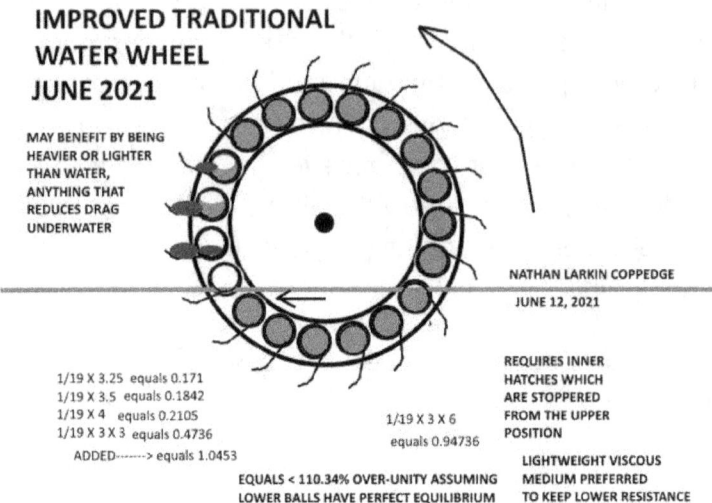

IMPROVED TRADITIONAL WATER WHEEL JUNE 2021

MAY BENEFIT BY BEING HEAVIER OR LIGHTER THAN WATER, ANYTHING THAT REDUCES DRAG UNDERWATER

NATHAN LARKIN COPPEDGE
JUNE 12, 2021

REQUIRES INNER HATCHES WHICH ARE STOPPERED FROM THE UPPER POSITION

LIGHTWEIGHT VISCOUS MEDIUM PREFERRED TO KEEP LOWER RESISTANCE LOWER

1/19 X 3.25 equals 0.171
1/19 X 3.5 equals 0.1842
1/19 X 4 equals 0.2105
1/19 X 3 X 3 equals 0.4736
ADDED--------> equals 1.0453

1/19 X 3 X 6 equals 0.94736

EQUALS < 110.34% OVER-UNITY ASSUMING LOWER BALLS HAVE PERFECT EQUILIBRIUM WITH ZERO FRICTION

APOLLO DEVICE

APOLLO DEVICE 2 Nathan Coppedge

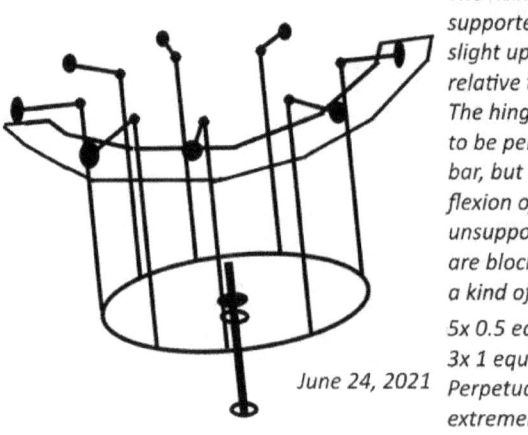

June 24, 2021

The rising weights are supported on a very slight upwards incline relative to the hinges. The hinges are blocked to be perpendicular to the bar, but allow upward flexion of the bar. The unsupported members are blocked, so provide a kind of leverage.

5x 0.5 equals 2.5
3x 1 equals 3
Perpetual motion, when extremely low friction!

It is found five arms may be ideal for this device (2:1.5 minimum mass advantage), and a fairly shallow angle, but with the track pressing upward very gradually to relieve some pressure on the rising levers. Original version of this, different from the original Apollo device, was with one arm and a counterweight, with the counterweight having a wheel, (2019/10/03)

...

CRESCENT DOMINOES

Possibly a good principle. I only regret that it looks like a phallus. Could you do a similar thing on a smaller scale? I have done a lot of experiments, and in similar mobile-weight based experiments I have noticed the return motion may not be exactly as much as the initial motion / energy. Some experiments seem to have good results, but sometimes very similar experiments have negative results. Maybe you could build a similar device using the crescent pendulum design on top of a row of dominoes, with the far right and far left domino having the crescent slope attachment and a small metal bead, and the entire pendulum being blocked at a shallow altitude. The motion of the dominoes might convert into the motion of the bead, which would then tilt the last domino back against the others.

—Response by Nathan Coppedge (July 2021) to video by Vikram
Gupta: https://www.youtube.com/watch?v=1raDr-qRtrg (https://www.youtube.com/watch?v=1raDr-qRtrg)

AVANT-GARDE ESCHER MACHINE

"Nice work, how long did it take to get the measurements right?" —Abdullah Khan

"I just had the same setup as a previous experiment with the Escher Machine. The height of the lever wasn't difficult, I had the lever and wheelie lying around also from another experiment. It took a little while to find the exact position for the wheel, had to line up the outer rim with the wire and the inner rim with the backboard. Hardest part was probably building the backboard and wire thing and positioning it at the correct angle. Definitely takes some experience. The combined X * Y angle is supposedly 1.1 degrees upwards sloped but I'm not sure, maybe more extreme in the case of the lever for some reason." —Nathan Coppedge

...

JULIOTONY'S ENERGY, "AIR FLOTATION DEVICE"

"I just wanted to comment that [it] might be improved by external cups which fill internal balloons so as to cause upward air motion which then gets repelled by the buoyancy of the air at the top. If the air could propel the conveyor, and be compressed by simple gears, there would be no need for an electric air compressor. [This could be the perfection of Frank Tatay's Grav-Buoy device]." —Nathan Coppedge, July 16,

2021 https://www.youtube.com/watch?v=YuWrpkUisu4 (https://www.youtube.com/watch?v=YuWrpkUisu4) (https://www.youtube.com/watch?v=YuWrpkUisu4 (https://www.youtube.com/watch?v=YuWrpkUisu4))

...

EXPONENTIALLY-EFFICIENT BALANCE: SMALLER BALL LIFTS HEAVIER BALL LONGER VERTICAL DISTANCE

"Normally this would mean that marble on the left would have to push with more force, if the force of the middle weight could lift the heavier marble without the smaller marble. In the case of a balance there is only return to an equilibrium, not heavy lifting" ---Nathan Coppedge, comment 2021–08–16

"Unless it's exponentially-efficient, the lever end will act exactly like a mass balancing with the opposite end of the balance, at least normally... However, this experiment of mine is not faked, and it does seem to still act as a lever somehow, which is interesting because the short end of the lever is gaining benefit from both the equilibrium on the other balance, and also the advantage of the lever by counterbalancing the force present in the lever. When slight additional mass is added to the lever end it can be considered as free lifting added to the heavy weight on the right, so long as the amount of mass without leverage included is compensated in the constant mass on the far left. Thus, adding a small amount of mass on the left will lift the mass on the far right including the free mass compensated already in the added leverage in the center. This apparatus will actually benefit from using high leverage in the middle, so long as the added mass on the lever is compensated by a counterweight in addition to the units of leverage in proportional mass units as compared to one unit of effective leverage." —Nathan Coppedge, comment August 18, 2021

"According to an approximation of the math, the mass on the right could be 11X, while the marble is 2X. The attached marbles on the left are 2X additional, while the attached marbles in the middle are 4X. If we analyze this system, what you are calling leverage in the middle is just leverage X 4. Meanwhile, the opposing force is a maximum of 1 X 4 plus 11 X 1. In terms of the maximum forces the difference of mass is 15:4 favoring the outside weights. The difference of leverage comparing inner to outer is about the same: 3:1 favoring the inner mass, or maybe 3:2 favoring the inner mass which would be worse for the effect of the 4X weight. The question is, when NORMALLY do we get 5.5X lifting power merely by comparing 15:4 with a 1:3 advantage? NEVER I GUESS. And the case does not get better if you think there is an advantage in the outside weights because of using two different 1X levers, because the outside weights already have a massive weight advantage, so the difficult part would be returning the 11X mass upwards, which means that the better advantage would be to give additional influence to the inside lever. So, I don't get the analogy when people say it's a typical lever. It doesn't seem to be. " —Nathan Coppedge, comment August 18, 2021

"All I should have to say is... Normally with a balance... If we compare 1:3 leverage to 3:1 mass it should not equal a 5X advantage. Normally when masses have the same effective leverage they will have to move longer distances to apply less mass. I suspect this is true whether the mass is being added or removed." — Nathan Coppedge, comment August 18, 2021

"If you try to do the same experiment yourself, you will find it's not as easy as it looks. It's not just because it's close to equilibrium, it is also because of the choices of ratios. I am particularly interested in how the ratios of mass and leverage interact." — Nathan Coppedge, comment August 18, 2021

"The point I was making was, if you can lift or drop a lighter mass to cause a heavier mass to rise or fall as a consequence, this means that you can cause elevators to move close to automatically simply by counterbalancing even a lighter falling weight against even a heavier rising weight. Since you can always raise or lower either weight in equilibrium, it ends up being a net advantage." —Nathan Coppedge, comment August 18, 2021

"It's easy to overlook this, but I will point out, unlike the two adults and a kid example, it's not just moving less distance, it's also moving FEWER DEGREES." —Nathan Coppedge, comment 2021–08–19

"This is without force inputted other than masses at rest. The [videos] are an application of this principle to lifting a much heavier mass with minimal input whatsoever. Since normally the miniscule additional input mass moves a slightly longer vertical distance, the implication here is that when the smaller mass moves a smaller vertical distance this can mean that the same counterweight which was perfectly efficient already can be even more efficient." —Nathan Coppedge, comment 2021–08–20

"Consider a balance with two equal heavy sides, and two equal dangling weights. You will find the two weights are not able to lift eachother, instead they tend to stay in equilibrium. You see, it is not the same case. In this case I just mentioned, the way to move the second weight would be to add additional mass to the first weight, or to move the first weight longer distance. But that is not what is happening. Instead, some special ratios and considerations have taken place." —Nathan Coppedge, comment 2021–08–20

"But if we say a child in an elevator can lift an adult in the same type of elevator with the child moving a shorter distance as well, we say that is ridiculous and against the laws of physics. But here it appears to be true, I argue for the first time. At least with the child with adults example the child would be moving the same number of degrees as either adult. But in this case, the 'child' is actually moving FEWER degrees AND is lighter. The difference in degrees might sometimes be 50% to 100% or more, and the difference in mass might be 5.5X or more. If you want loosy-floozy physics that still works, this is it essentially." —Nathan Coppedge, comment 2021–08–20

"So, if energy is based on mass at altitude, couldn't this be free energy? Because the smaller mass could be lifted by an even smaller mass, then be allowed to fall, thereby generating more than zero amount of force... As the force is simply equal to masses falling as you said..." —Nathan Coppedge, comment 2021–08–20

"To me, it seems like a subjective difference when you say it's a simple double lever. It's a miracle to me because I'm seeing a small marble lifting something 5.5 X it's mass with the smaller mass moving less distance. This is something I had not seen, or at least not directly, outside my experiments." —Nathan Coppedge, comment August 20, 2021

"If you analyze that a weight 5.5X larger is being moved a longer distance than a mass of 1X, that still seems interesting to me. I am not convinced it can be done with pulleys. The reason 'AGE 45' comes up is that has been noticed to be the time in life when most people stop thinking creatively. And at Age 36 I began seeing the beginning of it in myself, what is called 'hopeless conservatism'. I am very sensitive to this issue, and if someone doggedly defends science that in itself does not mean a whole lot if someone cannot think creatively enough to understand a new invention. What I see is more and more obfuscations. Maybe you understand my math, but I am not even sure of that. HERE IS WHAT I EMPHASIZE: While it is obviously true (since I know how the device works) that it is very similar to a balance, THERE IS ANOTHER NIFTY THING TO NOTICE which is that the second ball is 5.5. times heavier. AND... long pause... THERE IS A SECOND SECOND SECOND thing to notice, WHICH IS THAT THE FIRST BALL MOVES LESS LESS LESS DISTANCE." —Nathan Coppedge, comment August 22, 2021

"This is a development of an earlier deceptively less brilliant experiment in which a lighter ball was able to move a heavier ball longer distance using differences of angularity in support. Both were supported. The principle was then extended to moving a larger weight vertically without support for either weight." — Nathan Coppedge, comment August 23, 2021

"Whether it is over-unity seems to depend a lot on how much mass is being lifted, and whether the mass remains lifted if a portion of the mass is removed." — Nathan Coppedge, comment October 30, 2021

"If you measure to the middle of the mass on the left, which is where that big lump in the tape is, it's an approximation. I am not so much concerned about the exact amount of mass being added on the far left as long as I know that it is significantly less than the mass being lifted on the far right. When I was calculating it was one of my first attempts at this particular calculation, so my goal was to prove the principle, not match exactly with the experiment. The point is, both the experiment and the math work, and they are analogous. I think I was also trying to get the un-weighted resistance which wouldn't have included the marble on the far left, therefore the center of mass would have been very different than with the marble on the far left. In any case, the 1:3 measurement produces a conservative (in this case, higher) rating for the amount of mass necessary to lower the ball, meaning that the difference between masses is less than it might be. When the marble is removed, of course, if all is well and good the heavy marble will be lifted on the right in this setup, so that would only incorporate the measurement which includes no additional weight from the small marble on the far left." —Nathan Coppedge, comment October 30, 2021

"This particular device cannot easily become a perpetual motion machine because it does not use mostly horizontal motion in any of the directions, therefore, it is bound by some of the same constraints as a conventional vertical wheel, though it is still potentially useful because of its interesting properties. The advantage of horizontal motion is that altitude may be gained without opposing with more than around 60% mass. If combined with a device like the above, results might be seen with incredible gains in efficiency, though no perpetual motion device of that type has yet been designed that I have seen." —Nathan Coppedge, comment October 30, 2021

"Normally in a balance a mass must be opposed by a mass moving longer distance in order to compensate for moving a larger mass. However, this was not the case." —Nathan Coppedge, response to Leslie who was possibly an employee at Knowledge Generation Bureau, January 15, 2022

...

EXPONENTIALLY-EFFICIENT BALANCE 2

"In this arrangement the advantage is you can certainly lift the heavier weight using a smaller weight. The disadvantage is if you want to lift the counterweight again the part at least of the heavier weight which has moved upwards will have to move downwards again. The efficiency is that there is 'wiggle room'. It does not require exact balance. Some weight might move upwards for free." —Nathan Coppedge, Comment to "M.P." youtube September 14, 2021

"With this device, if you lift the heavier marble with the lighter marble, and then drop the heavy marble, then the drop of the heavy marble exceeds the force used to lift the smaller marble. However, there are other factors such as the additional counterweights present on the right and left. The neat thing is the smaller marble moved less distance, so in theory if you simply compare the smaller marble and the larger marble as inputs and outputs, then it is as if it is functioning more efficiently than a balance. Possibly since the larger marble can lift all the additional counterweights by itself, then the counterweights are offsetting leverage X mass, not mass alone, in which case there may be an efficiency principle." —Nathan Coppedge, response to question: Nathan Coppedge's answer to Can you get more power out of a falling object than it took to raise the object to begin with? (January 25, 2022)

DOMINOES OF INCREASING HEIGHTS

Experiment by Coppedge originally from 2009. Video may have been republished on a second youtube channel sometime later.

Work was done on this since 2009 by myself and at least one academic, however, evidence did not emerge until perhaps January 14, 2022 that there may have been a principle of over-unity involving potential energy. Here is the description which might indicate over-unity from 2022:

In 2009 several people including at least one academic and myself were experimenting with dominoes, and discovered that structural energy is not conserved. Although there is stored energy involved, the mechanical energy released by simply tilting one domino is much greater than the energy required to cause the tilting. This is more so the case going down stairs for example, where we can see additions of energy in spite of the fact that the next units required less energy to set up than the top domino. Though perhaps it is hard to conclude that any of these domino examples are over-unity, it leads to a theory of self-resetting dominoes and it is interesting to think about. Intuitively in the mind somehow it means over-unity. In principle if the dominoes were so light that all their energy came from downwards acceleration, then if the dominoes could be set up at virtually no cost, then this might be like perpetual motion if acceleration were high. perhaps this could be done by rotating lightweight counterbalanced magnets into place to increase acceleration. It is interesting to think about.

Consider the argument that if the higher units have more stored potential energy so would be harder to knock down, now if you can release greater potential energy by using only a smaller unit of potential energy, this at least shows that potential energy is increasing.

—Nathan Coppedge's answer to Whats the closest people ever got to overunity? (...)

UNNATURAL TORQUE DEVICE

"Let me reiterate the experiment is not faked. My experiments are as serious as it gets for the materials I have to work with. A list of all the absolute requirements that must be followed. It turns out they are all required, it was a bit of a lucky experiment: (1) Supports for fulcrum are about 2X back, 1X front, angled at 45 degrees downwards from fulcrum, (2) Fulcrum is supported by a vertical pin attached to a horizontally 45-degree angled fixture which can rotate 360 degrees within the fulcrum unless inhibited by the lever, (3) 45-degree angled fixture is also supported at close distance inside the length of the fulcrum bar by a fixed hole running parallel to the lower end hole on the fulcrum support bar. (4) The fulcrum support bar runs between the lower and upper ends beneath the 45-deg angled fixture and is NOT firmly connected to the fulcrum pin AND NOT firmly connected to the pin rotating sideways and supporting the fulcrum pin, the lever also rotates laterally at a specific angle due to the angle of the entire apparatus, (5) The upper end of the fulcrum supports is supported by over 0.5 inches maybe significantly more of additional height underneath, various angles should be tested to maximize utility, (6) The back end on the lower side is supported as shown by a different 0.25 in approx. additional height which may be very shallow, but seems to help, (7) Short end of lever is heavier, (8) Long end has small amount of weight, (9) Weight on short end is attached ABOVE the short end of the lever, and can be turned 45 degrees vertically towards the lower end to create motion, (10) Longer end is supported by very smooth bar at significantly lower altitude than the fulcrum, (11) The smooth bar is kept some distance inside the length of the long end and the lever itself is smooth and straight, (12) The smooth support bar is kept at a sufficiently shallow angle as to allow motion to take place. (13) It may help to rotate the entire device slightly counterclockwise from above to achieve the ideal lever position." —Comment on reddit 2022–01–24 mostly reposted from earlier circa 2020: Anyone believe this can be applied to designing self-powered VE Project style cars? The Experiment is real. This is reposted from my work on Quora, so just understand in this case it must remain free and non-proprietary. I have made this material called Unnatural Torque available many times...

"I just figured out it may work through exponential efficiency: in this case, a heavy counterweight at short distance versus a lighter weight applying leverage. This allows the torque to differentiate between the lever and the counterweight." —Separate comment by Nathan Coppedge same day on reddit, 2022–01–24

VARIATION ON A SIMPLISITIC DEVICE BY JASON THOMPSON:

Maybe on a smaller scale, if you make a spiral with a smaller ball guided by the spiral pushing a larger ball, the smaller ball could push the larger ball and perhaps if the larger ball had more momentum it would cause the smaller ball to jump at the beginning of the spiral resetting the cycle. Maybe this could work sometimes by gradual increase of momentum over time. However, this is not where my intuition normally goes on these types of devices. The typical case is if the larger ball will not gain momentum from the smaller ball unless it is lighter at every point. That might shoot down my proposal, which is in turn better than slowing the Earth's momentum. — Comment by Nathan Coppedge at youtube video by Jason Thompson

EFFICIENT MAGNETS:

"No, leverage versus a short-distance counterweight is more clever than Da Vinci if used on the horizontal with support during half the motion, with or without magnets. This is one of the reasons the above video is actually an okay attempt at perpetual motion. It's missing leverage and a means to reset the cycle." — Nathan Coppedge, July 14, 2022

"Each side would take turns losing altitude, either with a lever advantage or with a mass advantage. There would also be reduced resistance by supporting a rolling element with a slight upward slope around 58% to 65% effective mass on the gradient when moving horizontally. The fulcrum would have to pivot horizontally at an angle not vertically to have the effect." —Nathan Coppedge, July 14, 2022

COUNTERWEIGHTED AIR PRESSURE TO OPERATE UNBALANCED LEVERS:

Responding to a video at: https://www.youtube.com/watch?v=B_tyMXRjOE4 " Nice experiment, this has nothing to do with what you talked about. Here is a suggestion: you might try a much lower pressure arrangement with the pressure working on a horizontally-rotating wheel in which the pressure sends a lever outwards at a high point (it would be positioned on a very slightly slanted axis-- the axis itself would be mostly vertical). The vertical slant could be used to automatically return the lever inwards on what I imagining is the upper left side. Perhaps since the resistance in a horizontal wheel is about 50% for the sliding lever, maybe a magnet or air pressure could be used to move the lever inwards or outwards depending on what is preferred. Maybe someone in your situation could succeed with the unbalanced weight concept where the falling weight is somehow 50% heavier or what have you. I'm not sure magnets are required, I'm trying to work with the concept you already built or something similar. I believe in a lot of different concepts however yours is not the worst. You might also try counterweighted air pressure to send levers outwards." —Nathan Coppedge, message to Mike Beiler, 2022–09–05

TETHERED WEIGHTS

In standard vertical wheel: "You could try attaching string from lower left to upper right, attached off of the wheel, and only having a fulcrum hinge on the other side so the string doesn't tangle. Similar things have nearly worked for me with the sideways-swinging lever wheel, which has a similar principle of moving under certain conditions." —Comment at video posted March 22, 2019 https://www.youtube.com/watch?v=ezWQYCJpPI4

Sideways swinging lever wheel: Response to Sacreed One (Youtube channel. His question was: I built this the same and exact way and my question is how is this going to work by its self?): "My plan for this type of device is to use tethers (tied strings from outside the system, like high up or lower down on one side of the wheel so that it does not get tangled. In fact, lower down to the left acting to only bring the levers in half way at the furthest extreme may work to bring the levers in just below the middle right, which might be the only available strategy because it would allow the levers to extend below while still reducing resistance on one side. The goal is to reduce resistance enough that the levers apply enough pressure naturally simply by swinging out below level, which seems possible according to my experiment. Another method might be to deflect the levers inward using angled panels, in which case perhaps the horizontally-swinging advantage is the only thing that would make it work, as that strategy has been tried with the vertical-swinging variety. You can also try using eight levers instead of four, if you find evidence that it works okay with four). Make sure the rods swing out horizontally at the sides instead of up and down. Now figure out some way where you can prevent the rod from swinging out on one side, by tethering the end of the rod so it stays mostly on one side throughout the whole rotation. It could be a challenge to prevent the string from getting tangled, but if the rods only swing out on one side of the middle and are blocked internally at a slight inner angle, so that the rod swings out automatically when it is slightly below level on the falling side, then I think it might be workable. The video does not really cheat, except that I have artificially moved the rod inwards on the near side of the camera. I think this could be done with tethers if someone is very innovative. One of the secrets I found is that there may be a specific kind of tether that crosses over the middle just slightly from above (and not tethered in-line with the wheel at any location, but rather from only one side, with one string for each lever all the same length), so that the lever may swing out and still drop a certain distance, yet also swing inwards slightly when the lever starts to rotate to the other side. That is the secret. Under those conditions it seems to work. In any case, I hope it works. My experiment succeeded somewhat and I have described the parts that are missing. Also keep the string lightweight relative to the mass of the levers." —Youtube comment at the original sideways-swinging lever device video: https://www.youtube.com/watch?v=Y2gktCXrh2c&lc=UgwvV7qVW5mMthOLA6V4AaABAg 2023-05-22

SWIVEL LEVER W/ COUNTERWEIGHTED SPIRAL:

Inspired by video at: https://m.youtube.com/watch?v=EVJbQqleiJ8

"You can also try modifying the wheel to be horizontal with a peg, and counterweight in the wheel. This could give the swivel lever an advantage pushing on the spiral." ---Nathan Coppedge, comment at a VE Project video

ADVANCED BAILING METHOD:

One possible design I just thought of is to use a tilted horizontal water wheel, which pumps water up through the center, and then allows it to fall along the upper slant of the wheel. Maybe this is too reminiscent of the so-called Archimedean screw, but if the water is lifted mostly horizontally with a gradient lifting advantage, maybe the leverage advantage will cover other differences in some narrow cases. Good luck! Maybe an efficient method like gears can be found for lifting the water through the middle or something similar, or even something which rotates to allow the rising water to remain mostly horizontal.
—Response by Nathan Coppedge 2022–09–28 to video by VE Project
at: https://www.youtube.com/watch?v=99dYSsQ6YSY

BENNY WATER CUTTER:

Water cutter sounds like a waste of energy. What is your explanation for the extra energy? Mechanical? Surely you don't mean electric only? Sounds like you plugged it into the wall and forgot it was plugged in.

They know if you had a product you could sell it to someone local first. You'd be trying to set an affordable price unless you did something better than computers or cleaner than solar power.

ANACHRONISTIC GROUND TRANSPORT:

"There are details that are hard to explain to the skeptically-minded. In principle though, given sufficient midpoint height in the marble, after it gains altitude on the track it could lift a second counterweight by applying leverage again. It has momentum after all. It is a controversial principle." --- Nathan Coppedge, October 21, 2022

NATHAN COPPEDGE'S PRACTICAL PERPETUAL MOTION HANDBOOK

JER RAM'S ROULETTO:

4.5*1est*5*0.58
= 13.05 large wheel
1.5*1est*1*0.68
= 1.02 small wheel

13.05 / 1.02
1279% OU ???

August 19, 2022
Concept from 2009 - 2018 or earlier by Jer Ram

JER RAM'S ROULETTO
Diagram by Nathan Larkin Coppedge

"Jer Ram's perpetual motion machine remains ranked top amongst all the designs by about +1000% a number he wouldn't give me himself (his concept remains perhaps theoretical, I am not sure) even though he may not have been using the same math, or I may have borrowed my math from him." — Nathan Coppedge, 2023–05–05

...

ORFYRREUS' BESSLER WHEEL

There is a clever looking version of the Bessler Wheel, but I have heard it doesn't work, possibly (my own explanation) that the wheel is vertically oriented and thus resistance cannot be reduced easily. Spirals have their own problems as well though some spiral devices using external counterweights and a plunge lever have potential.

...

FERROFLUID DEVELOPMENTS

Something to think about is maybe 'magnetic water' this is probably not a very good concept, but it can inspire ideas similar to ferrofluid. Ferrofluid may use a lozenge or perhaps some type of magnetic membrane to lift the water perhaps at an angle, with the drop occurring at the same angle but supported only by the wheel or conveyor.

ADDITIONAL NOTES:

- Bedini by his reputation as a scammer is unlikely to have found a fully-working model, as his devices are most likely plugged in or rely on batteries. However, if he discovered how to use 0.5 mass * distance in his devices or if he discovered an over-unity computer method, these might have provided him a path to early perpetual motion. However, judging by what I can research on the web, bedini is most likely simply interested in building modified conventional motors. If he can prove otherwise it is taking a long time

- Stanley Meyer may have had a secret. It sounds a bit like a U.S. public relations stunt. In any case, extracting energy from water is not real free energy, and would have disastrous consequences in the long-term.
- Troy Hurtubise has an interesting general concept but it sounds to me like he is most interested in mech suit type thinking which while brilliant tends to be a bit exaggerated. If he has discovered a way to genuinely power a mech suit for free, then he is years ahead of his time in combining advanced technology and perpetual motion. An analysis of his name reveals that if the title is 'Troy Hurtubise Perpetual Motion' then the soul is: 'If you are fighting a romantic war and if when you are inventing perpetual motion you just give up, then it's not a fight with intellectuals or soldiers.' So, I would guess he is about impressing women. I would guess this puts him at a low technical level as far as new science as most women are traditionalists.
- Thomas Henry Moray is likely superstition, given how

emotionalized things were during that earlier period. However, perhaps he invented something like antigravity or electrostatic pressure. Maybe exciting to government but not necessarily the average person and probably outdated by now or hard to research.
- I am still researching the channel 'free energy bliss' but from what I gather so far, they can be quite clever but they are not focused enough on the right concepts to make something that would compete with solar or fusion. Perhaps they will remain creative and think of something maybe by stealing ideas from Nathan Coppedge.
- Perpetual Motion Lab is known for doing videos that are technical but not really on topic. Some of them may involve robots or weapons-type inventions, but they are not as meaningful for perpetual motion as they look at first, at least given the part of their content that is available to me.
- Perpetual Motion Junkie likely posts older video type content. I found the

quality of their work somewhat dirty, though I would be open to changing my mind if they present better content or if they had significant content I was not aware of.
- I don't remember 'Warehouse 13' very clearly, however, I think my impression was that they do ornate-looking models however they have not presented the best designs yet, though they get points for using spidery-looking and cyberpunk-style constructs.
- I believe what someone calls the 'perpetual motion channel' does not have a lot of content posted yet, so I'm not sure why [an AI] credits them as being one of the top channels. Maybe they have hidden content I haven't accessed yet? I could hope but I doubt it would be infinitely better...

[In 2023 the added notes] are a reasonable list of people I haven't researched yet if it turns out there is no one working on authentic perpetual motion yet who is very public and with good build skills.

Nathan Coppedge (b. 1982), perhaps best-known for his philosophy and perpetual motion machine designs & theory, is a philosopher, artist, inventor, and poet and member of the international honor society for philosophers. An abstract artist in Hyper-Cubism (sometimes credited to being worth $1,000,000 or more, there is now a book available) and philosophical writer, he has had work translated into Spanish, French, Italian, Portuguese, and Greek and has sold abstract Hyper-Cubism internationally. A one-time member of Tesla Society UK online and PESWiki, and founder of many Facebook groups, he lives near Yale University.